"This book should be required reading on every campus, regardless of size. Recent history has shown no school is immune to random violence. Carolyn Mears offers a guide to responding to the unthinkable; a practical map to shortening the road to recovery."

—Donald Donahue, Assistant Professor, University of Maryland; former Program Director, Health Policy & Preparedness, Potomac Institute for Policy Studies

"Carolyn Mears…provides an excellent firsthand analysis of the reactions and phases individuals experience on the pathway to recovery. She also provides a variety of practical and realistic activities to help with the recovery process."

—John Nicoletti, Police Psychologist, Nicoletti-Flater Associates

"This book has particular relevance to international schools because these schools are often located in unstable countries or places where the risks are higher for terrorism or catastrophic events. Equally important to taking preventative measures is to prepare for the aftermath of an unforeseen trauma or catastrophe. By presenting the varied real-life accounts this book provides a framework for every school administration to ask the questions that will help develop a plan that is best for each unique school."

—Katherine Johnson, Director, Human Resources, Singapore American School

# RECLAIMING SCHOOL IN THE AFTERMATH OF TRAUMA

## ADVICE BASED ON EXPERIENCE

EDITED BY
CAROLYN LUNSFORD MEARS

First published in 2012 by
PALGRAVE MACMILLAN®
in the United States—a division of St. Martin's Press LLC,
175 Fifth Avenue, New York, NY 10010.

Where this book is distributed in the UK, Europe and the rest of the world,
this is by Palgrave Macmillan, a division of Macmillan Publishers Limited,
registered in England, company number 785998, of Houndmills,
Basingstoke, Hampshire RG21 6XS.

Palgrave Macmillan is the global academic imprint of the above companies
and has companies and representatives throughout the world.

Palgrave® and Macmillan® are registered trademarks in the United States,
the United Kingdom, Europe and other countries.

ISBN 978-1-137-26854-9      ISBN 978-1-137-01082-7 (eBook)
DOI 10.1057/9781137010827

Library of Congress Cataloging-in-Publication Data

Reclaiming school in the aftermath of trauma : advice based on
experience / edited by Carolyn Lunsford Mears.
    p. cm.
    1. School crisis management—Case studies. I. Mears, Carolyn L.

LB2866.5.R43 2012
371.7'82—dc23                                    2011040186

A catalogue record of the book is available from the British Library.

Design by Newgen Imaging Systems (P) Ltd., Chennai, India.

First edition: April 2012

10 9 8 7 6 5 4 3 2 1


Transferred to Digital Printing in 2012

*We know the world by the stories we are told about it.*
*Denzin & Lincoln, 2005*

# CONTENTS

# ACKNOWLEDGMENTS

THE LIST OF INDIVIDUALS WHO HELPED IN THE CREATION of this book is long, and my gratitude to each is immense. To the writers, readers, reviewers, colleagues, interviewees, friends, and family, I feel so very fortunate to have your support and encouragement. I dedicate this volume to you and to others like you who care about people and want to help make their lives better.

Those who have transformed their own sorrow and loss into the service of others are truly inspirational, and I want to express my profound gratitude to them and especially to Gerda Weissman Klein and her late husband Kurt. Dearer souls never walked the planet.

Thank you, to all who agreed to be interviewed so others could learn from your experience—including Frank DeAngelis, Paula Reed, Bev Williams, Crystal Miller, Monette Park, Susan Peters, Mary Swanson, Barb Hirokawa, Ben Lausten, Kallie Leyba, Kiki Leyba, Krista Hanley, Michelle Wheeler, Jenn Smull, James Walpole, Bryan Krause, Ellen Stoddard-Keyes, Mary Pat Bowen, Karen Quiring, and many others who asked not to be named in print.

A special thank you to Mary Taylor and Monette Park who voluntarily took on the task of editing every chapter. Your careful reading and insightful suggestions were wonderful.

Thank you, chapter authors—you graciously responded to my request to share your experiences and insights for the purpose of helping others— even though initially to most of you I was a stranger. You have so vividly communicated the experience of trauma and the lessons that can be learned from it: Alan Kirk, Steve King, Paula Reed, Jethro Lieberman, Bridget Ramsey, James Whelan, Marilyn Saltzman, Nina Lyytinen, Kirsti Palonen, Russell Jones, Katharine Donlon, Kelly Burns, Kathryn Schwartz-Goel, Mary Kate Law, and Michael Dorn.

Thank you to those who supported this work in such a variety of ways— Marian Bussey, Bruce Uhrmacher, Susan Silver, Betsy Thompson, Larry Hincker, Nancy Feldman, Burke Gertenschlager and Kaylan Connally at

Palgrave Macmillan, and Deepa John at Newgen Knowledge Works. And, of course, my heartfelt gratitude to my incredible friends who have given ongoing moral support—Mary, Lynn, Sharon, Paula, Barbara, and the RCS.

I especially want to acknowledge the pride I feel for my sons, Austin and Brian, who had the courage to struggle in the aftermath and emerged stronger for it.

And, of course, my deepest appreciation for my loving, amazingly tolerant husband, Connally, who supported this journey every step of the way and has made my life beautiful by allowing me to share his.

# INTRODUCTION

THIS WAS A DIFFICULT BOOK TO WRITE, AND YOU need to know that it might be a difficult book to read. In the past, whenever I have spoken to groups about the aftermath of school shootings or other terrorizing events, I have given a similar caution, for the matters being addressed may tug at your own sense of well-being and cause you discomfort. How could they not?

If you are a teacher, administrator, or faculty member, you have prepared for your career by building content and process knowledge as well as expertise in methods for facilitating learning. The thought that you might be called on to lead a school or teach in an environment characterized by trauma is, of course, unsettling. It's always easier to avoid thinking of adversity, but denying risk doesn't make it go away.

The purpose of this book is to help you prepare for the worst that you can imagine, so if it occurs, you will find yourself able to make the kinds of decisions that need to be made. Whatever your role or relationship to schools, the experiences shared and the advice offered in this text can help you know what to expect and, if necessary, respond better.

I have long known that *disaster is hard,* but only as a sort of distant, intellectual truism. I have always empathized with individuals whose world was turned upside down by personal loss or large-scale catastrophe, but I never realized what those words meant until my community faced a disaster of its own, the rampage shooting at our local high school . . . Columbine.

Traumatic loss now was more than an abstraction. It had a name and a face, and it walked among us for many months—years, in fact. And in moments when the breeze is just right, the sky is the same shade of blue, and a helicopter hovers overhead, it can come roaring back with all its fury.

My younger son was a student at Columbine High School on April 20, 1999, and experienced the assault on his school firsthand. Later, when things were supposed to be getting "back to normal," I watched in abject helplessness as he, his schoolmates, and the entire school community struggled to find their way through what seemed to be a never-ending nightmare.

Knowing that inevitably tragedy would strike again, that somewhere another troubled youth would pick up a gun to kill his teacher, or a tornado would destroy a town, or a plane crash would claim lives, I wanted to find a way to help. I enrolled in the University of Denver in order to conduct dissertation research to learn what helped families recover after the Columbine shootings. Since completing that research, I have been contacted by educators in other communities who were trying to rebuild their world and reclaim their school after a tragedy. They asked what advice I could give as they attempted to teach traumatized students, and I realized that the need for information was growing.

"We know the world through the stories we are told about it" (Denzin & Lincoln, 2005, p. 641), and to help you know the world of trauma, this anthology offers stories written by individuals who have experienced devastating events and met the challenge of returning to teaching and learning in the aftermath. Their situations differed, but many of the problems they faced did not. While primarily addressing large-scale disasters, this book provides strategies that can also help students who struggle with their own personal tragedies. Assault, the sudden death of a parent, bullying, domestic violence, homelessness, the suicide of a friend...the list of traumatizing exposures is long. But whatever the source, trauma is trauma. It is very personal and very individual in its expression, regardless of its scale as viewed by others.

When I decided to compile this anthology, I asked the authors represented here to describe their personal experience related to a traumatic circumstance and then share what they thought was important for others to know about trauma and learning. Even though I was a stranger to all but one of the authors at the time, they readily agreed to contribute to the project, for, having gone through challenging times, they stood ready to assist others.

With any open-ended request such as this, responses will reveal what is of greatest significance or interest to the individual, and since people process events in very different and highly personal ways, authors addressed issues that were of greatest relevance to them. The result is a wide variety of perspectives and topics, all very instructive about the individual experience of trauma in a school setting and the process of recovery.

This text is, in a small way, testimony to transcendence and recovery. In it you will see the devastation that teachers, administrators, professors, counselors, students, and parents faced and the things they found beneficial in the aftermath. The significance of their insights made organizing this book particularly difficult, for it seemed that each chapter needed to be read first. However, that sort of holographic presentation is impossible in print, so I have adopted the following design.

**Section One: Understanding Trauma** provides a context for understanding traumatic experience, its consequences, and its implications for educators and students.

- In Chapter 1, I introduce the prevalence of traumatic exposure, its significance for educators, and an overview of trauma-related behaviors and learning difficulties that educators might observe.
- In Chapter 2, post-disaster intervention expert Alan Kirk and clinical social worker Steve King describe the biological changes in the brain that result from traumatic experience and the profound and long-lasting effect on cognitive and behavioral functioning.
- Chapter 3 brings the abstract term *trauma* to life through the personal narrative of Columbine teacher Paula Reed, who guided her students to safety during the attack and struggled as she returned to teaching in the aftermath. Her story exemplifies Kirk and King's description of traumatic stress and provides a view of one person's journey through tragedy and PTSD.

**Section Two: Learning from Trauma** provides insights and advice from a variety of circumstances. Each chapter offers a story of traumatic experience and the strategies that were developed to meet the changed environment and increased demands that resulted.

- In Chapter 4, Jethro K. Lieberman, professor of law, describes the upheaval experienced following the 9/11 terrorist attacks and the actions that were necessary to restore function to the New York Law School, eight blocks from the World Trade Center. In dealing with the crisis, the school had to address the traumatic effects of physical devastation and psychological horror compounded by the knowledge that the country had been attacked by a foreign enemy.
- In Chapter 5, Bridget D. Ramsey shares strategies she employed as principal of a New Orleans, inner-city public school that was converted to a charter school after Hurricane Katrina. The vastly altered needs of students meant that educators had to address personal, social, behavioral, and cognitive impairments of students, faculty, and families. Loss of infrastructure, services, social support networks, and basic necessities across the entire Gulf Coast region complicated attempts to reopen schools and resume teaching.
- In Chapter 6, science teacher James A. Whelan describes the logistics and adaptations that were involved in reopening a K-12 charter school for children of faculty and staff employed by Tulane and other private universities after Katrina. Applications for this school to operate as a charter had been submitted *before* the storm hit, which gave educators a head start in resuming school afterward. A partnership with area universities after the storm worked to everyone's advantage.

- Chapter 7, which gives an inside look at recovery from a school shooting, is based on my interviews with Columbine principal Frank DeAngelis, science teacher Bev Williams, and student Crystal Miller. This chapter includes a principal's reflections on leadership after a school tragedy, a classroom teacher's approach to meeting students' needs, and a student's life choices that helped her move toward recovery.

- Chapter 8, by Marilyn Saltzman, describes the ways that a rural community and school healed together after a drifter invaded Platte Canyon High School, took hostages, and murdered a student in Bailey, Colorado. Recovery there was enhanced by the strength of community, trusting relationships among service providers, clear rules regarding media access, and the powerful influence of the parents of the murdered child who turned their pain into efforts to assist students everywhere.

- In Chapter 9, psychologists Nina Lyytinen and Kirsti Palonen discuss the interventions and psychological aftercare that were provided for teachers and other school personnel following a rampage shooting in Jokela, Finland. Recommendations address advance planning, organization of services and interventions, provision of mental health intervention, media and communication, and helpful accommodations for educators.

- In Chapter 10, Russell T. Jones, professor of psychology at Virginia Tech, and several of his doctoral students summarize the mental health model and interventions that were employed in response to the campus shootings there. The authors review a three-tiered approach to counseling services, disaster mental health, the importance of culturally competent care, findings from the Virginia Tech investigation, and implications for educators called on to support mental health recovery.

- In Chapter 11, school safety expert Michael Dorn describes his experience as a victim of bullying and provides recommendations that schools can use to build a culture of respect and safety for all students. Of special note to schools is the importance of bystanders who speak out, the willingness of adults to take action, and strategies for creating safe school environments everywhere.

- In Chapter 12, I share advice from students, teachers, parents, clergy, counselors, and others about the challenges they faced after school-associated trauma and the responses they considered beneficial as well as those that they thought harmful.

**Section Three: Putting Pain to Work** builds on lessons learned from the experiences provided in the previous chapters, offering a synthesis of overarching considerations and resources to help prepare for the aftermath.

- Chapter 13 suggests common themes from the lessons learned.
- Chapter 14 provides an annotated list of readily available materials, tools, and resources, most of which are free.

I hope you never need the information in this book. But if tragedy ever strikes where you live, or you ever wonder how to help a single student or staff member who exhibits signs of trauma, you will have the benefit of advice from people who have faced the challenge of reclaiming school and the educational process in the aftermath.

### REFERENCES

Denzin, N. K., & Lincoln, Y. S. (2005). Methods of collecting and analyzing empirical materials. In Denzin, N. K., & Lincoln, Y. S. (Eds.), *The Sage handbook of qualitative research* (3rd ed., pp. 641–49). Thousand Oaks, CA: Sage Publications.

# SECTION ONE

# UNDERSTANDING TRAUMA

# CHAPTER 1

# TRAUMA COMES TO SCHOOL

## CAROLYN LUNSFORD MEARS

> April 21, 1999: I can't believe we have a meeting this morning. I can't believe there is electricity in my apartment or that water still flows out of my faucet. I peel apart the window blinds; I can't believe the neighbor is walking his dog.
>
> Kiki Leyba (2004)
> Teacher, Columbine High School

A SUBURBAN COLORADO HIGH SCHOOL IS RIPPED APART BY gunfire and bombs as angry students embark on a murderous rampage. On the plains of Kansas, a tornado siren screams, shattering the peace of a summer afternoon and sending residents in a frantic rush to safety. At a seaside resort, the National Guard rumbles in to enforce mandatory evacuations as a Category 4 hurricane barrels toward the coast. Moments of chaos, fear, destruction—the sudden rending of the normal—leave the everyday almost unrecognizable and the future in doubt. Routine events, once so easily taken for granted, now appear strange and incomprehensible.

Between one heartbeat and the next, the world is transformed. Worldviews are shattered; plans are crushed beneath the reality of the moment, and it becomes difficult to imagine what the future will hold.

A life-threatening or life-altering event violates the everyday and can generate long-lingering challenges. In addition to those personally exposed to the threat, effects of trauma are often felt by family members and others who witness or hear of the heartbreak. Even if injury and loss are averted, warnings of impending disaster can trigger fear and uncertainty, especially among children and youth.

That no-man's-land between event and resolution is the *aftermath*—the point where the pains of experience are still raw and angry, before the healing has taken hold, before the world begins to look livable again. Often a traumatic event commands front-page/prime-time coverage, and then slips from sight as other events move into the spotlight. The situation hasn't been resolved; it's merely been replaced by something more immediate.

## SIGNIFICANCE FOR SCHOOLS

Hardships posed by disaster or traumatic loss persist. Lives are upended, security is threatened, and all are forced to find ways to adapt to a new sense of "normal." Traumatic situations range from collective, massive upheaval of warfare or a terrorist attack to solitary personal experience of victimization or sudden loss. Some crises strike suddenly, but others are more subtle and ongoing. These may be suffered in the silence of unrelieved anxiety and fear. Students whose parents are in the military, for example, often experience stress from knowing that parents have been—or may be—deployed to a war zone. Living in a high-crime area creates an elevated baseline of traumatic exposure. Children of undocumented workers may live in fear that their family will be deported. A quiet residential neighborhood, without identifiable sources of threat, may be the scene of recurring child abuse. For those living with chronic anxiety, any subsequent disruption may elevate the level of traumatization. Whether an event terrorizes an entire school population or only a single individual, educators need to be alert to stress responses and understand what they can do to help from an educational perspective.

Psychologists, social workers, and counselors are taught to assess the impact of traumatization and to provide treatment that helps victims move toward recovery. Educators, however, are generally not given any preparation for working with students who are suffering from traumatic stress. In the event of a disaster or serious loss, most schools and universities will bring in counselors to attend to immediate psychological concerns. For individual cases, educators know procedures for referring the student for special services. However, while mental health professionals may be available to assess and counsel, *educators* are rarely taught how to adjust their instructional practice to meet the special needs that arise.

While it may be comforting to believe that bad things only happen to other people, "Trauma is a universal experience. It is no respecter of rich or poor, of profession or occupation, of country of origin or family of origin, of talent or personal purpose" (Bussey & Wise, 2008, p. 3). And, while it is nice to think that we can protect children from harm, the National Child Traumatic Stress Network reports that one in four school students in the United States experiences a traumatic event that can profoundly affect behavior and learning (2008).

Whether the trauma is a single event or ongoing, large-scale or small, caused by humans or the result of natural forces, the effect can be the same—students impaired by experience. The special needs that trauma causes, thus, have serious implications for teachers and faculty. Only by becoming aware of what trauma looks like in a classroom and how adjustments can be made to instruction, curriculum, and environments, will educators be able to help students resume their learning.

In this volume, you will find factual accounts written by individuals who have encountered trauma and are sharing their experiences to help others understand what it means for teaching and learning. In addition to their accounts, I would like to offer you the following fictionalized narrative, a composite created from a variety of sources, to illustrate how trauma from witnessing the tragic circumstances of others might be seen in a classroom.

Robert had always been a little small for his age, yet his winning smile and quiet sense of humor won him many friends. His teachers remarked about what a good student he was, seemingly able to master most subjects with ease and always willing to lend a hand if classmates needed help. When he was in 9th grade, he tried out for the baseball team and earned a spot on the junior varsity squad. He was sure he would make varsity the following year and committed to long hours in the weight room and on the running track to build his strength and stamina.

In the summer of his 9th grade year, a massive tornado struck Robert's hometown. There was considerable damage to the northern suburbs—buildings collapsed, the roof was ripped off the high school, trees were uprooted, several people were injured, and one was killed. Robert was out of town when the disaster hit—on a camping trip with his father and younger brother. They heard reports of the storm and immediately headed home. When they arrived, they were stunned to see the destruction on the north side of town, but the area where Robert's family lived suffered little damage. They considered themselves fortunate that the storm track had spared their immediate community.

By September, most of the debris had been cleared away, and though structures in the affected areas still showed signs of damage, daily routines began to take on a degree of normalcy, especially on Robert's side of town. His high school reopened on schedule, and students appeared eager to reconnect with friends. Students from the damaged high school to the north were not so lucky. Their school was unusable, and they were dispersed to other schools until repairs could be made. About a third of the student body was bused to Robert's school.

In spite of the disruptions, Robert seemed to do well that year, and with extra effort, he was able to make the varsity team. The following year, students who had transferred to Robert's school were able to return to their own school, and things seemed to be getting back to normal. However, Robert began to show increasing disinterest in school and seemed to be withdrawing from his friends. His growing irritability and moodiness caused many arguments at home. Although his parents were troubled by the changes in his behavior, they assured themselves that he was merely going through a stage.

At midterm conferences, Robert's parents learned that his teachers were concerned about his behavior, poor performance on tests, and failure to turn in assignments. Teachers observed that he chose to sit in the back of the room, no longer paying attention in class. When called on, he just glared in stony silence. By semester break, matters were worse. Robert had quit the baseball team, was skipping classes, and in danger of being suspended for disruptive behavior. All were concerned about the change in this formerly conscientious student. One teacher questioned whether these changes might be related to the tornado. She had noticed that Robert was not the only one who appeared to be struggling.

While Robert is a fictional character, this scenario exemplifies the situation of countless people worldwide. In this case, it was a tornado that shattered Robert's basic sense of safety and caused anxious concern. With his community not directly hit by the storm, it is easy to see how his stress was overlooked. However, this type of vicarious trauma, caused by witnessing the suffering of others, is not uncommon. For some students, even watching media coverage of natural disasters, terrorist attacks, or violent crime can elevate apprehension to a level that affects school performance. It is important to remember that disruptive response to trauma does not require direct victimization, nor does it quickly resolve itself.

## EFFECTS ON BEHAVIOR AND LEARNING

Traumatic response is not a matter of choice. It is neither a product of simple emotion nor an isolated psychological symptom. It has a biological basis, and while the brain and body deal with physiological changes (see Chapter 2), the traumatized individual is also struggling with complex feelings of fear, bereavement, and remorse. Intense grief may be felt for those killed or injured, for the loss of personal possessions and homes, for the loss of possibility, and for the loss of a worldview free from the knowledge of vulnerability and risk.

Psychologists use the term *assumptive world* to describe the "strongly held set of assumptions about the world and the self which is...used as a means of recognizing, planning, and acting" (Janoff-Bulman, 1992, p. 5). While our assumptions about life may not always be accurate, they provide a means of orienting ourselves to the world in which we live. Experiencing violation or witnessing a disaster may shatter these basic assumptions, and as a result, new assumptions and a tenable worldview must be built to integrate a new awareness that the world can be a risky place.

Accompanying the loss of the assumptive world is a psychological phenomenon known as the development of a *trauma membrane* (Lindy, 1985). Those sharing the experience of a traumatic event tend to withdraw and close ranks to avoid being misunderstood or further violated. This disconnection from others is created when people who are affected by a traumatic event feel distanced from those who have not lived the event and therefore cannot possibly understand a new reality that words cannot describe.

Survivors of a tragic situation often feel powerless. While individuals vary, these feelings can lead to an aggressive drive to assert control or, alternately, a sense of resignation. In cases involving loss of life, survivors often question their self-worth and may feel guilty for having survived while others did not. Some who are exposed to trauma suffer clinical depression; some exhibit increased recklessness, a greater potential for substance abuse, and other high-risk behaviors. Certain factors influence the severity and duration of the response, including such variables as the specific nature of and exposure to the event, witnessing extreme distress of others, cultural background, availability of resources, social support, and any pre-existing problems that are only made worse by the exposure. In general, recovery from acts of violence perpetrated by a classmate or a known individual is more complicated than recovery from natural disaster, since the situation raises feelings of betrayal and questions of trust.

Schools and universities are the hub of their surrounding communities, and any disaster that happens in the community is echoed in the educational setting, interrupting school routine and disrupting learning. For children and adults alike, the traumatic experience can be life changing. "Some of the . . . potentially devastating costs of disasters for children include missed school and reduced academic functioning; missed social opportunities; and increased exposure to life stressors, such as family illness, divorce, family violence, and substance abuse" (Silverman & La Greca, 2002, p. 13).

Trauma takes a toll on quality of life, on relationships, and on personal growth and development. For students, "a decline in school performance is one of the big responses to trauma, though teachers don't always know to look for this . . . [R]esearch from both Oklahoma City and 9/11 shows that almost a year later, children reported high levels of distress around family safety and difficulty focusing on schoolwork" (Shorr, 2006, pp. 21–22).

Parents who participated in my own research into the effects of the Columbine tragedy (2005) reported changes in their children's behavior.

For some, the effects were short-lived, for others more long-lasting and possibly lifelong. Responses that caused concern included the following:

- depression
- anxiety and hypervigilance
- loss of sense of safety
- anger, irritability, rage
- oppositional behavior
- withdrawal from activities and friends
- moodiness
- risk-taking
- sensitivity to exploitation
- sense of a foreshortened future
- academic difficulties
- memory loss
- difficulty expressing self
- disruption of home life
- difficulty sleeping

Parents and teachers also reported that they felt some of the same responses compounded by other concerns. They expressed uncertainty about how to parent/teach a child who had experienced "a war zone." Some described marital discord that ended in divorce. Any such secondary stressor exacerbated the psychological, behavioral, emotional, and cognitive impacts from the initial trauma. In the classroom, these impacts may be seen in

- impaired concentration
- impaired ability in problem solving
- problems with short- and long-term memory
- withdrawal from prior relationships and routines
                              (U.S. Department of Education, 2005)

The consequence of these symptoms is that a student's overall school performance can be impaired, resulting in

- lower grade point average
- increased number of school absences
- increased likelihood of dropping out
- increase in suspensions and/or expulsion
- decreased reading ability
- increased emotional upset
                    (National Child Traumatic Stress Network, 2008)

When presented with lists of discrete symptoms neatly laid out on the page, it is easy to see how trauma can disrupt the learning process, especially if a recent disaster has struck a community. It becomes more difficult for teachers and parents of a student like Robert, who did not personally experience loss, to connect school problems to a disaster, especially if the student did not exhibit any troubling symptoms early on. It also becomes more complicated to recognize traumatic effect when working with an individual suffering from chronic abuse, victimization, or bullying. However, when teachers or faculty observe significant changes in a student's overall behavior, it is important to assess the possibility that the changes are the by-product of a traumatic circumstance. It is also important to know what options are available to assist the student if this is found to be the case.

## PREPARING FOR THE POSSIBILITY

Mounting evidence points to the varied ways that trauma can affect learning, and educators need to know what they can do to help. While institutions and agencies plan for crisis prevention and incident management, insufficient attention is given to preparing educators and communities to assist their students in the aftermath of an event that has not been—or cannot be—prevented. Risk prevention is indeed preferable to managing recovery, and yet life is a risk and tragedies do happen.

Whether seen in broadcast images of destroyed communities or in the silent eyes of individual pain and abuse, an experience of trauma will affect students who sit in the classroom and the teachers who try to teach. Yet, the situation is far from hopeless, for many of the strategies that facilitate learning in the aftermath are actually extensions of practices that supported learning before. A caring teacher and a classroom that feels safe, with an established sense of order and rules that are consistently and fairly applied, give students the comfort of predictability. After a tragedy, returning to these routines can help restore equilibrium. Instruction that addresses the individual student's learning needs, cognitive strengths, and behavioral patterns facilitates achievement of curricular goals and standards. A trusting relationship and clear communication supports advancement. By building on these foundations, connecting the student with available mental health resources, and adapting to new needs that emerge, educators can assist students in continuing their academic progress.

Practices that were in place before disaster contribute to recovery. Resuming functional and organizational continuity supported by existing positive relationships provides a basis for moving forward. The *continuity principle* (Omer & Alon, 1994) emphasizes the importance of reconnecting

to existing resources and social networks in promoting reconciliation of life
before and life after a traumatic event:

> The more an intervention is built on the [student's] existing individual,
> familial, organization, and communal (e.g., schools, neighborhood support
> services) strengths and resources, the more effective it will be in counteract-
> ing the disruptive effects of [trauma]. Every available material, every person,
> and every event can become "therapeutic."...Intervention methods should
> be swift and simple as possible, and they should have a clear goal of normal-
> ization of stress reactions. (Klingman, 2002, p. 367)

After a tragedy, a return to school provides structure, a sense of order, and
an opportunity for social support that comes from interaction with peers.
However, teaching in the aftermath is not simply returning to business as
usual. While teachers should not be turned into therapists, they do need to
know about trauma and the implications for instructional practice. In some
cases, educators themselves have experienced the tragic event and need to
address their own response to the trauma.

Recognizing that there is no single *right* answer or approach to resuming
the business of school, the suggestions in the following chapters should be
viewed as starting points for discussions among administrators, teachers, faculty, students in licensure programs, crisis responders, counselors, and policy
makers. Understanding what will be needed—before tragedy strikes—will
increase the likelihood that all can work together to assess potential need,
develop viable strategies, and identify possible resources that can be called
on. The sort of collaborative networks and established communication
among providers that are required in the aftermath characterize a healthy
environment for schools and communities, regardless of traumatic exposure.
Whether reclaiming a school, or a university, or a classroom experience for a
single student, preparation improves the chances for recovery.

## REFERENCES

Bussey, M., & Wise, J. B. (Eds.). (2008). *Trauma transformed: An empowerment response*. New York: Columbia University Press.

Janoff-Bulman, R. (1992). *Shattered assumptions: Towards a new psychology of trauma*. New York: The Free Press.

Klingman, A. (2002). Children under stress of war. In A. M. LaGreca, W. K. Silverman, E. M. Vernberg, & M. C. Roberts (Eds.), *Helping children cope with disasters and terrorism* (pp. 359–380). Washington, DC: American Psychological Association.

Leyba, K. (Fall 2004). The lesson I learned. *Teaching Tolerance, 26*. Retrieved April 10, 2011, from http://www.tolerance.org/supplement/lesson-i-learned.

Lindy, J. D. (1985). Trauma membrane and other clinical concepts derived from psychotherapeutic work with survivors of natural disasters. *Psychiatric Annals 15*(3), 153–160.

Mears, C. L. (2005). Experiences of Columbine parents: Finding a way to tomorrow. (Doctoral dissertation). University of Denver. ProQuest UMI: AAT 3161558.

National Child Traumatic Stress Network. (October 2008). *Child trauma toolkit for educators.* Los Angeles. Retrieved July 14, 2011, from http://www.nctsnet.org/content/defining-trauma-and-child-traumatic-stress.

Omer, H., & Alon, N. (1994). The continuity principle: A unified approach to disaster and trauma. *American Journal of Community Psychology, 22,* 273–283.

Shorr, P. W. (Fall 2006). The longest days: Leadership during rebuilding and recovery from a disaster, *Threshold,* 20–23. Retrieved August 25, 2009, from www.ciconline.org.

Silverman, W. K., & La Greca, A. M. (2002). Children experiencing disasters: Definitions, reactions, and predictors of outcomes. In A. M. La Greca, W. K. Silverman, E. M. Vernberg, & M. C. Roberts (Eds.), *Helping children cope with disasters and terrorism* (pp. 11–33). Washington, DC: American Psychological Association.

U.S. Department of Education. (2005). *Tips for helping students recovering from traumatic events.* Washington, DC.

## WHAT'S NEXT?

In the following chapter, "Trauma's Effect on the Brain: An Overview for Educators," Alan Kirk and Steve King explain what happens to the body during a traumatic experience. The authors, who are experts in trauma response and post-disaster counseling, clarify the biological responses to trauma, including changes in the physical structures and functioning of the brain. As they show, recovering from trauma is not a matter of willpower or a choice to "just get over it and move on."

CHAPTER 2

# TRAUMA'S EFFECT ON THE BRAIN: AN OVERVIEW FOR EDUCATORS

ALAN KIRK & STEVE KING

March 21, 2005—a day like most others on the Red Lake Reservation, Minnesota, home to approximately 5,000 members of the Red Lake Band of Chippewa. At the high school, the Red Lake Warriors basketball team had just completed its winning season. Baseball season would begin in two weeks. Students were making preparations for prom, spring break, and generally enjoying the first signs of spring. This was a day like all others. Nothing special. Until...the incident.

It began around noon when a 16-year-old student shot and killed his grandfather and his grandfather's girlfriend. The student then drove to his high school and entered through the main entrance, where he was met by two unarmed security guards manning the school's metal detector. The student shot and killed one of the guards, as the other managed to lead a group of students to safety. The young gunman proceeded down the hallway and began firing into an English classroom. There he killed a teacher and three students and wounded three other students. A 16-year-old sophomore trying to halt the attack wrestled with the attacker and stabbed him with a pencil. His actions allowed students to flee the classroom. The price for this rescue was high; the young student was shot twice, leaving him in serious condition. The attacker continued his rampage, killing two more students and wounding two others before police could arrive.

In a shoot-out with the police, the attacker sustained gunshot wounds to the abdomen and right arm and retreated to a vacant classroom where he turned the gun on himself and committed suicide. In the brief though deadly attack, the troubled student killed a total of nine people, including a teacher, a security guard, and five students, and wounded eighteen others. The home of the Red Lake Warriors was transformed by this event from a high school to a community of survivors. The lives of students and educators at the school as well as those of members of the community at large would never be the same. Before March 21, safety at this high school had been assumed. *Trauma* and *hurt* were defined in different terms and with different understanding. Repercussions of this tragedy are likely to be felt at Red Lake for years.

## THE INCIDENCE OF LETHAL SCHOOL VIOLENCE

It is a common misconception that school-associated deaths in the United States are both random and rare. Highly publicized massacres such as the incidents at Red Lake, Columbine, and Virginia Tech have heightened the awareness of school violence, but most experts assert that such rampages remain infrequent. Yet, during the first decade of the millennium, 65 schools in 39 states experienced life-threatening attacks. In the years 1999–2009, a total of 284 deaths were reported in association with K-12 schools (National School Safety and Security Services, 2011). For the years 2007–2009, there were 149 deaths associated with public and private colleges due to homicide or manslaughter (U.S. Department of Education, 2011). Deadly attacks are seen across a variety of demographics and settings, and at all levels of schooling—primary, secondary, and college/university. Although rare considering the number of schools and students in this country, these losses point to a pressing social problem with a significant impact on the educational community. Rarely a week goes by without a report of a lethal shooting in or near a school.

In addition to losses from school violence, educational settings are subject to significant disruption by other causes as well. Natural disasters, community violence, suicides, fatal accidents, and a wide variety of life-threatening events bring tragedies that affect the educational arena. The gravity of the situation can only be understood by recognizing that the actual number of individuals affected by these events includes not only the victims and their family and friends, but also the witnesses to the traumatic loss, and often the entire school community.

## TRAUMA AS A MEMORY

In the past, clinicians and educators conceptualized memory and experience in rather vague terms, often defining the noun with itself. For example, a *memory* is often defined as something remembered. Likewise, an *experience* is conceptualized as something experienced. These definitions may be adequate for casual discussion, but to really understand learning, one must understand how the human brain processes stimuli from the outside world, how this information is stored in the brain, and how experience becomes a memory.

The human brain is an amazing organ. Its three pounds of specialized tissue can store a lifetime of images. The brain system consists of more than a billion neurons. Each of these cells has over 1,000 connections to other neurons, yielding more than 1 trillion neural connections. Recent research posits that the brain has an overall storage capacity of more than 10 terabytes—that's 10,000,000,000,000 bytes of information! (One megabyte equals roughly 1,000,000 bytes; 1 gigabyte equals 1,000 megabytes; and 1 terabyte equals 1,000 gigabytes.) Numbers like these are difficult to comprehend, so here's an analogy: If your brain could be connected to a television monitor, it would have the capacity to store 3 million hours of television programming. That's over 340 years of continuous playing!

Recent research has demonstrated that stored memory is persistent and enduring. Even though an individual may not be able to consciously recall a specific memory, memories remain chemically stored in the brain. This characteristic of memory has strong implications for those in positions to help individuals exposed to a traumatic situation. In fact, some researchers have likened the effects of traumatic experience to a sort of brain damage. For many survivors, the chemistry of the brain is permanently altered, giving rise to problematic behaviors and decreasing ability in certain key areas, learning being one of these key areas.

We have all seen how the human body heals itself after an injury or illness. Cuts, blisters, and scrapes transform into scabs that may fade slowly or may remain as permanent scars. The body's remarkable ability to heal, sometimes requiring the aid of medication or treatment, demonstrates its automatic efforts to recover from any loss of function due to an injury or physical trauma. Various environmental factors can influence the speed and quality of recovery, and even when the repair is complete we may find ourselves somewhat different from what we were before our injury. If you break your arm, for example, you may not recover its full strength or flexibility right away. In fact, you may be more protective of it even after the bone has healed, for the body remembers the experience even after the physical injury has healed. Similarly, exposure to a traumatic situation leaves an individual more vigilant for potential trauma and more likely to experience a traumatic response in the future.

Just as physical injury or disease taxes the body's ability to recover, so too does exposure to traumatic events or circumstance. Especially for children and youth, psychological trauma can have a profound and long-lasting impact even if it leaves few visible scars. Specific disruptive and painful events combine with an individual's unique developmental challenges and strengths to determine the severity and duration of aftereffects. Threatening events and behaviors are processed differently by individuals of different ages and levels of development. Adolescents and adults, for example, may be alarmed by dangerous weather but genuinely terrified by the sound of gunshots. Younger children, unfamiliar with the context of these sounds, may be more frightened by the crashing fury of a storm than by the potentially more lethal threat of gunfire. The degree of powerlessness felt by individuals in the face of various dangers influences the level of psychological trauma they experience. For example, the startling sounds of thunder during a violent storm may overwhelm a child, while older individuals in the same storm may feel less frightened and more in control, since they know of ways to avoid injury. However, their knowledge of the random and unpredictable lethality of gunfire alerts them to the terrifying nature of the situation and their vulnerability in it.

Traumatic experiences create structural changes in the human brain that can alter mood, cognitive ability, and behavior patterns (Nadel & Jacobs, 1998). These biological changes are the result of the body's overall capacity to heal itself and to learn from experience, especially if it has significance for survival. The painful and frightening experience of extreme circumstance such as touching a hot stove creates unique neurological pathways and patterns that are stored in the brain as memories. These memories serve a survival purpose, since they can help that person make protective decisions if exposed to similar events in the future. Learning to fear certain people, places, or circumstances can be an adaptive skill that helps us avoid injury for a lifetime. The neurobiological changes associated with these traumatic memories function to calm the fears a person experiences and ultimately to restore emotional equilibrium and stability (Perry & Pollard, 1998). Through this process, an individual may learn to respect and safely manage fire, and while the resulting cautious behavior may seem extreme or unnecessary to others, the neurologically permanent nature of the "burn from the stove" memory works to ensure avoidance of another burn experience.

These familiar examples lay the groundwork for a deeper discussion of the nature of trauma and the neurobiological changes that it brings. An elemental understanding of how trauma changes the brain can help guide educators who must plan and enhance instructional strategies to meet the needs of students who have been exposed to violence and traumatic events. It also has implications for educators about how to shape learning environments to meet the needs of *all* students.

## THE SIGNIFICANCE OF BRAIN DEVELOPMENT AND THE EXPERIENCE OF TRAUMA

The human brain develops in a complex, sequential fashion beginning with the brain stem, which is the part of the brain responsible for heart and lung functions that are critical to the survival of the organism. Given this essential role, it is clear that the development of the brain stem must be virtually complete for an infant to survive life outside of the womb. Any sort of prenatal injury or insult to the fetus, such as alcohol or drug use by the mother or domestic violence perpetrated by the father, can compromise brain development and functioning throughout the child's life. With a snowball effect, prenatal insult can result in cognitive developmental delays, poor emotional control, conduct problems, dysfunctional peer relationships, and school failure (Inbinder, 2002). Trauma experienced prior to birth can create neural pathways that sensitize and predispose the developing brain so that certain environmental stressors are experienced as traumatic, while other children may have greater resiliency to the exact same stressors (Perry, 2009). This can, in part, begin to explain different responses to abuse and neglect experienced by siblings from the same family.

The sequential nature of brain development and its adaptive and malleable qualities ensure that a child's brain develops abilities that are best suited to cope with its immediate environment. However, this ordered process results in children being vulnerable to diverse physiological effects of trauma depending on their age and brain development at the time of their traumatic experience. For example, healthy, higher level brain functioning (e.g., abstract thought) depends on the successful development of earlier systems that control more basic functions such as emotional regulation or concrete problem-solving skills (Perry, 2009). If brain development is disrupted by trauma, the child will neurologically adapt to account for or compensate for an abusive, toxic, or traumatic environment. The memories and lessons learned (often subconsciously) can profoundly alter a child's emotional and social functioning as well as the ability to benefit from future learning experiences and environments (Corbin, 2007; Perry, Pollard, Blakley, Baker, & Vigilante, 1995).

The unique nature of the brain is that it can heal itself from experiences that have no direct connection with the central nervous system. Thus, as a child's brain processes traumatic psychological or emotional experiences as *injuries*, the brain mends or rewires itself in response to traumatic occurrence.

Most of us have traumatic experiences of one sort or another in the course of our life, and we carry memories and even emotions or behaviors associated with these experiences with us as we age. Fears or anxieties connected to stormy weather, loud voices, vicious animals, or other distressing situations are common, and many of us can trace a personal behavior of today back to an unpleasant or painful experience as a child. Even the smell of a specific aftershave or cologne can bring back feelings of sorrow due to the loss of a favorite family member associated with that fragrance. While the loss may not be traumatic in nature, the stimulation of the senses activates neural pathways that have been specifically influenced by painful loss or unpleasant events from the past, and as a result, the original feelings of loss are to an extent remembered, relived, and re-experienced.

This process, which is not pathological in and of itself, demonstrates how the brain mediates life experiences into memory (Perry & Pollard, 1998). Dysfunction develops when environmental stressors are *extreme* and/or *persistent* for an individual who lacks the resources to adequately cope and who is thus overwhelmed by the trauma. Incidents of physical abuse, sexual assault, a persistently violent home or community, and exposure to natural disasters and warfare all inflict environmentally induced trauma that activates special biological and psychological adaptive mechanisms in an effort to ensure the survival of the individual.

In response to an extreme trauma, the central nervous system enlists a collection of unique processes that seek to repair the damage inflicted by the experience. These include increased intensity of electrical activity through complex patterns of neurons in the brain. Unique chemicals and hormones are released in an effort to repair and return the traumatized individual to a state of emotional balance. If a trauma is severe enough, memory created before the trauma is rewritten into new neural pathways that can't return the system to its previous level of functioning. Surges of biological defenses that respond to abandonment, abuse, and terror form new pathways to incorporate the experience and in the process create a new but traumatic memory.

To illustrate this, consider children who may have an age-appropriate fear of the dark but, after being subjected to repeated sexual abuse late at night, build traumatic memory that stimulates a fear so strong that as adults they need to sleep with all the lights in the house turned on. This seemingly extreme behavior is the result of the brain's best efforts to recover from trauma and to avoid similar trauma in the future. It is easy to anticipate the inevitable consequences of this adaptive behavior on interpersonal relationships. How we manage our emotions, interact with peers, and establish intimate relationships all hinge on this process and the coping memories and mechanisms we have acquired throughout our development.

## THE STRESS-RESPONSE CONTINUUM:
## HYPERAROUSAL AND DISSOCIATION

As the human species evolved, successful navigation of an often hostile environment was essential for individual and family group survival. Over the passage of time, two distinct but related behavioral responses to extreme and/ or threatening stressors developed: (1) an active and aggressive hyperarousal response, and (2) a passive, withdrawing response known as dissociation.

*Hyperarousal,* which is often known as the "fight or flight" response, involves the body's preparations to defend itself. Heart rate and blood pressure increase, and muscles expand, preparing the individual to fight for survival or to flee toward safety. Glucose levels in the blood rise and provide an instant source of fuel for energy, activity, and action. This hyperarousal response, while available to both men and women, is seen more frequently in males and is considered adaptive in men from an evolutionary perspective. In order to protect family members from predators (human or otherwise), this aggressive/active response functioned to preserve the family unit. If successful, this response capacity would be genetically passed on to one's offspring (Perry & Pollard, 1998; Perry et al., 1995).

Boys who have been traumatized are likely to react to perceived social and environmental stressors with the hyperarousal response. Impulsivity, angry outbursts, and aggressive behavior in a school-age boy may signify a history of abuse or trauma that has neurologically generalized and sensitized him to unconsciously activate this behavioral response pattern to seemingly mild or innocuous events. For example, if "Jim" is late arriving to class, telling him to take his seat may result in an angry and disruptive outburst that is disproportionate to the situation, thus leaving his teacher and classmates confused and even frightened by his extreme response. If Jim is already plagued by the fear of failure, exposure to peer ridicule, or disapproval from authority, his response may be barely contained and ready to activate at the slightest perceived threat. As a result of prior trauma, he has been sensitized to overreact in order to protect himself from further insult or injury (Gill, 2010; Perry & Hambrick, 2008).

We know that trauma can hardwire the brain to be extraordinarily sensitive to stress or stimuli that most people cope with as a matter of course (Nadel & Jacobs, 1998). Attempts by the central nervous system to return an individual to equilibrium create traumatic memories that can lead to maladaptive response patterns, generating defensive outbursts and acting out behavior designed to stabilize the system (Perry et al., 1995; Perry & Pollard, 1998). Unfortunately, organized social structures like schools, sports teams, and social groups or clubs, by their nature, can inadvertently stimulate stress and anxiety with which traumatized children have little capacity to cope. These children may eventually become unwelcome in these settings and gravitate to peer groups with the lowest threshold for membership and the highest level of tolerance for antisocial behavior.

At the other end of the response continuum, the passive response of *dissociation* can be pictured as a mirror image of the hyperarousal stress response. The traumatized individual retreats or "surrenders" in response to violent or frightening environmental threats and stress. During a dissociative episode, physical activity is dramatically decreased and the individual "withdraws into a shell." The dissociative process is marked by a slowing of the heartbeat and a decrease in blood pressure as the body's blood vessels expand. Dissociative states are characterized by avoidance behavior, emotional numbing, and a blank facial expression. In school or other social situations these can closely resemble prosocial behaviors such as compliance and/or resilience on the child's part. It can be easy to overlook a conscientious student like "Margaret" who, after being physically abused as a child, seems to be "just fine." The abuse may not seem to have affected her at all; yet the dissociative coping response, in the service of personal survival, may be masking an underlying turmoil that without attention can increase the risk of long-term psychological trauma (Gill, 2010).

If a child adopts this response to early trauma, seemingly benign situations and normal stressors may stimulate dissociative behavior later in life. This coping strategy disadvantages the child because important internal coping skills are not learned, and emotional adaptation and growth are stunted.

Unconscious reliance on the dissociative behavior to cope with stress also inhibits the development of adaptive thinking skills. Individuals who unknowingly dissociate when faced with adversity or pain may have limited experience and practice in thinking through stressful social situations and personal problems. They may lack the interpersonal skills and judgment needed to avoid situations later in life (e.g., abusive or exploitive relationships) resulting in painful and dysfunctional relationship patterns (Gill, 2010).

## ACUTE TRAUMA-RELATED STRESS REACTIONS

Traumatic experiences vary considerably in terms of severity, duration, and the level of danger or threat perceived by the victim. While chronic child abuse takes a profound and long-term emotional and behavioral toll on children, a wide variety of traumatic events can trigger dysfunctional stress-related responses in individuals of all ages. For example, surviving a serious automobile accident, being raped, or witnessing a violent attack on another may occur as discrete, single incidents, but the same adaptive and reparative neurological processes are activated by the brain in an effort to rebalance the traumatized system. Acute Stress Disorder (ASD) is a cluster of emotional and behavioral reactions to a traumatic event that are evident immediately after the event occurs. In many ways, ASD is similar to the more familiar post-traumatic stress disorder (PTSD) but is shorter in duration (American Psychiatric Association, 2000). ASD symptoms have been observed in people who have survived natural disasters, serious accidents, violent assault, armed robbery, and in people who have witnessed community violence such as workplace or school shootings.

## CHILDREN

Studies of acute stress reactions in children and adolescents who have experienced or witnessed a violent and traumatic event have identified individual, victim-personality characteristics that can inhibit recovery. For example, previously traumatized children within a larger group of children exposed to a subsequent traumatic event will be more likely to experience the most severe symptom clusters (Salmon, Sinclair, & Bryant, 2007). These individuals have already been sensitized to trauma and have neurological memory patterns that have been activated by the newest traumatic event. Dysfunctional coping strategies already in place will be stimulated and put into motion. It is not a matter of choice; the response has been learned as a survival mechanism, and the distress is quite real.

Individuals characterized by deep doubts about their own abilities to cope with problems or stress of any sort can have severe reactions to a troubling event and require longer periods of time to recover and come to terms with what they have experienced. Those who hold a predominantly negative view of the world and expect painful and frightening events to be a regular or normal part of everyday life will struggle to regain prior levels of functioning after a traumatic experience. People who, in general, lack trust in others or have little confidence in the security of the world around them can experience a persistence of traumatic symptoms and feelings. Children with negative appraisals of the capacity of others to help them will have additional difficulty in the healing process as well (Salmon et al., 2007). These characteristics make the prognosis for complete recovery from trauma guarded for these children. Youth with greater self-esteem, a stronger sense of self-efficacy, and more trust in others may be better able to withstand the longer-term consequences of trauma survival with lower levels of emotional disturbance and interpersonal dysfunction.

## ADOLESCENTS

Empirical research into the effects of trauma on adolescents has primarily focused on stress reactions to an isolated type of trauma (e.g., sexual abuse survivors or accident victims), and as a result little is known about the struggles that victims of different types of traumatic events may have in common or whether their traumatic stress responses may be event specific. Saul, Grant, and Carter (2008) surveyed 1,581 individuals between the ages of 12–17 who had survived a traumatic event during which they experienced a strong fear of serious injury or death. For the purposes of their study, the researchers categorized traumatic events in the following ways. Teens were asked if they had been a witness to violence (i.e., an assault, a shooting, or a stabbing); had been sexually or physically abused; had experienced a serious accident; or had been exposed to a natural disaster. Study participants were then asked to indicate whether they had experienced one or more of a series of trauma-related symptoms, including difficulty sleeping, intrusive memories of the event (flashbacks), irritability, emotional numbing, physical or psychological stress when reminded of the traumatic event, hypervigilance, exaggerated startle response, and so forth. The purpose of the study was to determine if the current understanding of how teens cope with the aftereffects of trauma can be generalized to other age groups or if youth trauma symptomology has a wider variance depending on specific youth or specific incident type variables (Saul et al., 2008).

In this study, females reported experiencing more stress-related response symptoms than males, and older teens reported more disruptive and dysfunctional symptoms than younger study participants. More severe and problematic traumatic stress responses were reported by youth who had been exposed to violent trauma (assault) as opposed to those exposed to nonviolent trauma (natural disaster). All of the youth surveyed tended to identify with a majority of symptoms that fit within a dissociative type response pattern (Saul et al., 2008). This finding may be due in part to the few acting out or hyperarousal-type behaviors that the researchers included in their survey. In other words, the hyperarousal symptoms often seen in a fight-or-flight response are not currently universally understood to be indicative of traumatic responses and are underrepresented in children identified as suffering from traumatic stress.

This situation brings into question some of the current ways we understand how traumatic stress is experienced and how it is expressed by children and adolescents. The hyperarousal spectrum of response patterns may be underrepresented in present models of traumatic stress symptomology. They do not take into account disruptive, defiant, and oppositional behavior patterns, which are then, unfortunately, attributed to issues other than trauma survival such as conduct disorders, attention deficit hyperactivity disorder (ADHD), impulse control problems, and learning disorders (Perry & Hambrick, 2008; Saul et al., 2008).

## COLLEGE-AGE STUDENTS

By the time a student reaches age 18 or 19 (i.e., college age), there is greater likelihood that the individual has either witnessed or personally experienced some sort of traumatic event. As a result, any further traumatic exposure has the potential for producing responses that reflect the cumulative stress. In addition, any trauma-related response will be played out in a very different environment on a college campus than it would in a K-12 school. Indeed, the physical structure, social culture, and the nature of faculty-student relationships on a college campus are quite different from those found in K-12 schools.

In the relatively closed system of primary and secondary education, students are under the care of teachers who occupy the role of in loco parentis, with rights and responsibilities for looking after students who are under their supervision. The authority granted to educators through school and community policy allows teachers to make decisions and mediate students' disruptions and acting-out behavior in the classroom. If a student's attitude or performance is troubling, for example, the teacher is empowered to address those issues in school and, if the situation merits, to contact the parent or guardian and develop a shared approach to the problem. If a student's behavior appears symptomatic of trauma-related stress, educators can meet with the parents, and if abuse or neglect is suspected, child protection laws require a referral for further investigation.

College faculty have no such relationship or access to their students' parents or family members. On a college campus, students are treated as adults. By privacy laws and institutional policy, faculty are prohibited from sharing information about their students—even their grades—unless the student authorizes the release of that information in writing.

Faculty and staff may find themselves working with individual students who demonstrate behaviors symptomatic of trauma response, such as declining grades, poor attendance, inattention, disruptive behavior, or substance abuse. These situations could indicate a need for counseling or health-related support; yet communicating with the student may be difficult, since he or she, as an adult, has freedom of choice, and faculty lack access to family members who might be able to assist.

In our increasingly complex society, older students require new social and technical skills to successfully transition to adulthood. Older adolescents/young adults (age 17–21) face daunting developmental challenges as well, for they are still growing physically, emotionally, socially, and neurologically. Many parents witness with surprise the return of their college student who has grown two inches by the end of the school year and who has become concerned with relationships, politics, and world events. These changes are evidence of the ongoing development of this age group.

While the transition to independence is an individual journey for each young adult, all face a number of significant challenges. The emergence and exploration of issues such as sexuality, increased freedom of choice, substance use, and financial independence can create significant stressors for college-age students. In some cases, the onset of mental health issues, such as depression and schizophrenia not previously experienced can, in part, be associated with the acute pressures unique to this age group. Separation from familiar resources including parents, friends, and teachers places additional demands that many are not ready to deal with.

Times of transition from secondary education to the demands of college, the workplace, or the military are difficult hurdles for young adults to navigate. This population remains at risk for the neurobiological consequences of traumatic events, and the instability and their changing developmental needs make the role of post-secondary faculty and staff particularly challenging.

## IMPLICATIONS FOR EDUCATORS

Few professionals touch the lives of as many individuals as do teachers. Few witness the emotional and behavioral consequences of the important life events, whether they are uplifting or traumatic. School is often the first place where traumatized students' problems and struggles are manifested. Educators are positioned to observe and identify potentially traumatized students and to recognize their learning needs. When multiple students or an entire school community have been traumatized, an appropriate and forward-thinking response by the administration and faculty, inside and outside the classroom, can facilitate the healing process and help the school community move toward recovery.

When teachers become concerned about a recently traumatized student in their classroom, they need to know a variety of practical actions that they can take to assist the individual, as well as to minimize the potential impact on the class as a whole. While a teacher is not a therapist, individual educational adjustments or modifications can help that student move toward achievement of educational goals as well as promote a healthy and supportive environment at school.

The crucial first step is to become alert to the emergence of academic or behavior problems and trauma-related response patterns in a troubled student. If a formerly quiet and well-mannered student becomes defensive and hostile about missed assignments or poor attendance, a variety of causes should be considered, including the possibility of a recent traumatizing experience. If trauma is identified as the cause of the behavioral change, this student may need referral for outside medical or therapeutic interventions, and the teacher may need to adjust academic assignments to accommodate the student's short-term learning limitations. Problematic stress response behaviors, like impaired concentration, intrusive memories, or unexpected dissociative "daydreaming," can dissipate with time and appropriate care, and concerned teachers can facilitate this process.

If the student's disruptive behaviors or academic problems need to be confronted, it is important to assess when and where to do so. A student who formerly could take gentle correction or counsel about poor performance on a test may misperceive a teacher's concern as disapproval or threat when none is intended. While a student recovering from trauma needs time to devote to the healing process, the structure and predictability of the school day can be beneficial. Returning to a workable routine often provides a measure of comfort and a sense of normalcy. Since some traumatized individuals experience deep guilt and shame associated with their experience, educators and administrators need to be alert to issues of confidentiality and privacy. It can be quite challenging to balance a student's needs for individual consideration and his or her desire to return to "normal" as quickly as possible so as not stand out from the crowd.

Many children will confide in teachers they trust about the details of painful and disturbing experiences they have had. While educators and school staff should be vigilant for any situation requiring reporting of potential abuse, they should realize that these efforts are potentially growth promoting for the student and signal efforts to heal from the experience. Honest expressions of empathy, compassion, and care can mean a great deal, and the process of confiding in an adult who is not a parent or designated as a mental health professional can bring a sense of balance and acceptance from the "real world" that all trauma survivors hope for.

Many of the approaches that assist a traumatized individual can be applied to a classroom or school community that has been collectively traumatized. School communities that have experienced devastating natural disasters or violent outbursts such as suicide clusters, fatal accidents, acts of terrorism, or school shootings undergo a traumatic response reorganization and healing process similar to that of traumatized individuals. A school community that has been collectively traumatized will struggle to achieve equilibrium as a group in addition to the trauma-response behavior patterns they will be experiencing individually. Routine academic functions that demand extra levels of concentration and attention to detail, such as high-stakes standardized tests or placement exams, may need to be rescheduled or postponed to account for the cognitive disruptions associated with trauma exposure. Administration, faculty, and staff will ultimately need to find the delicate but essential balance between "business as usual" and the unpredictable collective healing agenda.

In addition to the potential impact of trauma on classroom instruction, educators should recognize that school traditions, celebrations, and rituals, which can help restore a positive collective identity, will take on new meaning in the aftermath of a collective loss or trauma. Each event needs to be considered in light of the changed circumstances, for what may have been appropriate before may need to be re-envisioned or modified to account for the new situation. Whether the disaster is of human origin or a natural phenomenon, any traumatic event that strikes the school campus itself can create levels of group survivor guilt, anger, and fear that may cast a pall over all school functions, including sporting events, artistic performances, club competitions, dances, graduation, and other public displays of school community pride and cohesion. The senior prom, for example, will take on new, painful significance for many students if several classmates died earlier in the year in a drive-by shooting or fatal bus crash. Collective myths and "stories" will develop around the memories of violent events and the people lost as a result of them. These are efforts to shape the meaning of what the community has experienced into something that can be explained and understood, as unfathomable as the trauma may be.

Educators can profoundly empower their students by serving as shepherds to this process. By understanding how trauma affects individual and collective systems, the faculty and staff can model healthy response behavior and normalize the healing process by validating and reassuring students that their thoughts and feelings are not indicative of permanent disability but rather steps along a path toward healing and recovery. Teachers, who themselves may need accommodations as they return to the classroom, can serve as witnesses and guides to the long way back from traumatic injury for their students. Educators can provide a safe place to express painful feelings and rework memories.

Equally as important to the recovery process is the promotion and creation of a well-structured and goal-oriented environment that expects and encourages success within an atmosphere of acceptance and flexibility. One of the first casualties of traumatic experience is a sense of control over one's life. Formerly independent and confident students who have been exposed to trauma have their developing sense of identity and mastery of the world pulled out from under them. The structure of the school day provides hourly opportunities for these students to regain a sense of stability and predictability to their emotional world. Compassionate teachers who help facilitate a process of academic and social successes for these students can contribute significantly to their healing and long-term recovery.

When a large-scale crisis strikes a college campus, the road to recovery can be especially challenging, for the diversity of students and faculty makes it difficult to meet the variety of needs that will arise. Institutions of higher education are, by design, home to a much more diverse population than characterizes a typical elementary or secondary school. Often students and faculty have come from other locations; indeed many come from other countries, bringing a wealth of diverse cultures and worldviews. In routine situations, this rich diversity creates the milieu for an energetic and positive sharing of ideas and information. During a crisis, however, it can create added concerns, especially for students who, because of geographic distance, may lack the reassuring support from family and friends "back home." Administrators and campus leaders who nurture a healthy sense of the institution as a caring community, in advance of traumatic occurrence, can foster an environment that embraces and supports diversity of culture and traditions, including those around grief and mourning.

One of the most immediate needs when returning to a school after a large-scale event is to ensure that students feel safe enough to continue with their education. While primary or secondary schools are restricted communities with only a few points of entry, most college campuses are quite open, with numerous points of access and extended hours of operation. In addition, traumatized students with impaired behavioral controls, aggression, or anger may present a risk to teachers who are charged with managing students within the classroom. Faculty walk a fine line between controlling and motivating a student, while acknowledging that the issues may originate with the trauma that has been experienced. College students, as adults, have increased access to weapons through various outlets. Whereas a high school or middle school child would have more difficulty in acquiring weapons, college students have easy access to gun sales. Parental safeguards for weapon storage often do not apply to a college cohort. At the same time, while attending to students who may demonstrate behaviors characteristic of hyperarousal, faculty should not overlook the needs of students who may have withdrawn into a dissociative state.

## IN CONCLUSION

Communities are always at risk for experiencing loss due to a natural disaster or other circumstance. Increased awareness of school-associated violence during the last two decades has forever changed our perception of safety in K-12 schools and on college campuses. Moreover, the increasing incidence of personal abuse and neglect experienced by individual students further complicates demands on educators. Yet, in spite of the potential for experiencing traumatic loss, the human brain is an amazingly resilient organ that can adapt to changes. This chapter has considered some of the recent research in neurobiology of trauma, and this basic foundation can assist educators as they attempt to create learning situations that benefit all their students, including those who are survivors of traumatic events, abuse, or neglect.

## REFERENCES

American Psychiatric Association. (2000). *Diagnostic and statistical manual of mental disorders,* 4th edition. Washington, DC: Author.

Corbin, J. R. (2007). Reactive attachment disorder: A biopsychosocial disturbance of attachment. *Child and Adolescent Social Work Journal, 24,* 539–552.

Gill, S. (2010). The therapist as psychobiological regulator: Dissociation, affect attunement and clinical process. *Clinical Social Work Journal, 38,* 260–268.

Inbinder, F. C. (2002). Torrential tears: The relationship between memory development, early trauma, and dysfunctional behavior. *Clinical Social Work Journal, 30*(4), 343–357.

Nadel, L., & Jacobs, W. J. (1998). Traumatic memory is special. *Current Directions in Psychological Science, 7*(5), 154–156.

National School Safety and Security Services. (2011). School-associated violent deaths and school shootings. Retrieved March 13, 2011, from http://www. schoolsecurity.org/trends/school_violence.html.

Perry, B. D. (2009). Examining child maltreatment through a neurodevelopmental lens: Clinical applications of the neurosequential model of therapeutics. *Journal of Loss and Trauma, 14,* 240–255.

Perry, B. D., Pollard, R. A., Blakley, T. L., Baker, W. L., & Vigilante, D. (1995). Childhood trauma, the neurobiology of adaptation, and "use-dependent" development of the brain: How "states become traits." *Infant Mental Health Journal, 16*(4), 271–291.

Perry, B. D., & Hambrick, E. P. (2008). The neurosequential model of therapeutics. *Reclaiming Children and Youth, 17*(3), 38–43.

Perry, B. D., & Pollard, R. (1998). Homeostasis, stress, trauma, and adaptation: A neurodevelopmental view of childhood trauma. *Child and Adolescent Psychiatric Clinics of North America, 7*(1), 33–51.

Salmon, K., Sinclair, E., & Bryant, R. A. (2007). The role of maladaptive appraisals in child acute stress reactions. *British Journal of Clinical Psychology, 46,* 203–210.

Saul, A. L., Grant, K. E., & Carter, J. S. (2008). Post-traumatic reactions in adolescents: How well do the DSM-IV PTSD criteria fit the real life experience of trauma exposed youth? *Journal of Abnormal Child Psychology, 36,* 915–925.

U.S. Department of Education. (2011). *Summary campus crime and security statistics.* Retrieved March 13, 2011, from http://www2.ed.gov/admins/lead/safety/crime/criminaloffenses/index.html.

\* \* \*

## CONTRIBUTOR NOTES

**Alan Kirk, Ph.D.,** professor of social work at Kennesaw State University, is an internationally recognized expert in post-disaster intervention, psychological trauma, and post-traumatic stress. His professional practice background includes service with the

U.S. Air Force, the Veteran's Administration, and full-time private practice in the Miami/Fort Lauderdale metropolitan area. Dr. Kirk has served on international and national disaster intervention teams related to natural and man-made disasters, including the tsunami disaster in Southeast Asia, Hurricane Katrina, Hurricane Andrew, and the attack on the World Trade Center.

**Steve King, Ph.D.,** assistant professor of social work and human services at Kennesaw State University, is a clinical social worker with over 20 years of experience working with children and families in a wide variety of clinical settings. He has particular interest and experience in the development of integrated communities of care to meet the needs of multiproblem children being served by mental health, education, and juvenile justice systems.

**WHAT'S NEXT?**

Chapter 3 provides a living example of the neurological and behavioral consequences of extreme trauma. In her personal narrative, Columbine English teacher Paula Reed shares how it *feels* to have post-traumatic stress disorder, how her trauma affected her family, and how she adjusted her teaching to meet the world that had changed around her.

CHAPTER 3

# BARELY FUNCTIONING: THE EXPERIENCE OF PTSD

PAULA REED

*First, my story—the answer to the perennial question, "Where were you that day?"*

I WAS TEACHING A VOCATIONAL ENGLISH CLASS, AND HALF of my students were in the computer lab with my teammate, working on a project. I had stayed in the classroom with kids who had already finished it and were catching up on other assignments. One student was reading a book on serial killers, and we joked about the fact that people who knew them always described serial killers as being "so quiet."

Did I think for a moment about a quiet sophomore I had taught two years before and who was now a senior? Not at all. When I heard the name Dylan Klebold later that day, I didn't recognize it. It wasn't until I saw his picture on the news in the wee hours of Wednesday morning that I realized I knew him.

Anyway, the kids were on-task, and I had to go to the bathroom, which was just across the hall, so I ducked out for a few minutes. When the fire alarm went off, I thought it was a drill, and I followed my kids into the beautiful spring day. When kids ran by me screaming, "They've got guns!" I thought they were overreacting to something—a rumor of some kind. Kids did not bring guns to Columbine. When I heard gunfire, I thought there was a real fire after all and something was exploding. When administrators ran by and told us to get the kids over the chain-link fence and away from the building, I did what I was told, calmly certain that everything was going to be just fine.

As I walked to the community library (not to be confused with the school library) in the warm sunshine with hundreds of kids, some asked whether they could get their cars and go home. I said no, because I was sure they'd get this whole thing cleared up and we'd finish the school day. Queen of Denial, that's me.

The library was in chaos. My colleagues and I wandered from group to group, calming kids down, helping them locate siblings and call their parents, who then streamed into the community library and a nearby elementary school looking for their children. We had no way of knowing who had gotten out of the building and who was still trapped inside; it was pure pandemonium. I spent God knows how long getting kids to sign sheets of paper saying they were there, letting parents examine the lists, and then trying to get the information faxed over to the elementary school for the panicked families there.

After a while, it occurred to me that I should call my husband. It hadn't dawned on me that we were a major news story. I wasn't outside the school library. I didn't see gravely injured Patrick Ireland propel himself out of the window to rescuers below. I didn't know what was going on back at my building, and by the time I talked to my husband, the event had been on television and radio for over two hours. He'd been worried sick, but I had no sense of time. It felt like it had been 20 minutes at the most.

My father and his wife knew I was all right because they'd heard me on television. A local station asked me to get on the phone and report how Columbine parents could find their children. I did so, thinking it was a local broadcast, but it was actually CNN. I found that out much later. That broadcast also made it possible for my son's second-grade teacher to pull him aside after recess and say, "Your mom is all right, but something bad happened at her school." At some point, my dad called my mom, who was living in Florida and was frantic.

By about 4:00 in the afternoon, victims' advocates from the sheriff's department had taken over the jobs I'd been doing, so I listened to the news inside the library. At that point, the media was reporting 35 confirmed dead. (I've learned a lot about the accuracy—or lack thereof—of the news.) My bubble of denial burst. *Thirty-five*, I thought. *I must know some of them.*

At last I went to the local elementary school where the remaining unconnected children and parents were gathering. It was there that countless students told me they'd been evacuated past the body of Rachel Scott, a girl on my speech team with whom I was close. I looked at the first boy who told me that—her varsity mentor on the team, my state poetry champion—and I said, "That's not true. You're lying." Then I walked away.

I'm not proud of that. Another teacher snapped me back to my senses and told me that I'd better get back there and apologize and help that kid out, which I did.

In the elementary school gymnasium, another member of my team pointed out that no one had heard from Dan Mauser. Bob Curnow, a friend and regular judge at speech competitions, awaited word from his son Steve. Bob and I spoke lightly of how much trouble Steve would be in for not calling, for letting his family worry. It was so much easier to pretend that he was going to call at any minute, but the hours kept crawling past and we heard nothing.

Several sources told me that my colleague, Dave Sanders, had been shot and was being transported to a hospital, but for some reason, they would not remove his name from the list of missing people. He remained on that list, along with Rachel Scott, Dan Mauser, Steve Curnow, Isaiah Shoels (a former student of mine), Lauren Townsend (a senior that I'd taught as a freshman), and a number of other names unfamiliar to me then, but now etched in my memory forever.

By 7:30 p.m., I ran into Ron, my old principal. He'd been working in central administration and was privy to more information than most of us. I said, "Ron, when will we know? I can't leave without knowing where my kids are."

He said, "Go home, Paula. They're not releasing any names until tomorrow afternoon at the earliest."

I said, "But how do I know who's OK?"

He said, "There are as many bodies in that building as there are families in this gym right now."

Oh.

I don't remember exactly what I thought or felt. I said, "What about Dave? Everyone knows he was transported, but they won't take him off the list."

Ron said, "He wasn't transported. He's still in the building."

"What! But he was shot. Why is he still there?"

Ron paused. It must have been so hard for him to say, "There wasn't any need to take him out."

Oh.

A few minutes later a man stood up at the microphone on the stage in the gym and said, "Those of you parents who have not heard from your children, could you please contact your dentists for records?"

A woman went outside and threw up.

I went home and watched the news. That's when I saw Dylan and realized who he was—my former student.

## HOW COULD I TEACH?

A few days after the shootings, I went on *Nightline* with one of my speech students and several other kids. Ted Koppel had arranged a sort of video town meeting between us and a group from Jonesboro, Arkansas. They had been through a school shooting a year earlier. When I had a chance to ask a question, I asked the one that was most important to me at the time: How did the teachers there do it? How did they go back to school and teach? How could I help my students deal with what had happened?

This was my biggest fear. I knew I had to go back and teach, but I didn't have the first idea how. I was in such emotional upheaval, and from what I'd seen of the students at various gatherings, they were in no better shape. I just wanted someone to tell me that it could be done, because before we went back to school, it just looked impossible to me.

No one from Jonesboro had an answer for me. I don't remember what was said in response, but several people there pointed out later that I hadn't gotten an answer. I even received a letter from a woman who had watched the show and noted the same thing. She said she had an answer for me. "Be honest about yourself to the kids. Tell them when you're hurting."

I'm writing this reflection and hoping no one else ever has to ask this question, but giving my answer all the same. First and foremost, it can be done; you can do it. When the chips are down and you walk into that classroom, you will be what you have always been—a teacher.

I really didn't know where to begin. Everything I needed, textbooks, my grade book, my materials, kids' ungraded assignments, had been left behind when the fire alarm went off on April 20. Columbine High School was a crime scene, and everything there was being guarded night and day by the FBI and the Jefferson County Sheriff's Office. We weren't going to be able to return to our school until after police completed the investigation and repairs were made to the building. We would have to finish the school year at nearby Chatfield High. Two weeks out, we all felt a driving need for something normal. We desperately needed to *do* school, but really, how could I teach, and how could students learn?

The first thing I learned about PTSD was that it left you with the attention span of a gnat. Jeffco Mental Health tried to help with this. They came up with canned lesson plans that were all about processing and healing, but they were flying in the dark as much as we were. No one had had to do this before. Even the teachers and kids of Westside Middle School in Jonesboro had been able to go back to their own building, resume their routines. Most of us looked over the lesson plans and decided that they just didn't reflect our jobs as teachers, and despite everything, that's what we still were— teachers. Besides, if we all followed the lesson plans, kids would do the exact same activity all day long, class after class, utterly wallowing in the trauma. That didn't seem like a good idea.

My colleagues in the English department talked about it. Money had been made available to purchase books, whatever books we wanted to teach. A million or so copies of *Chicken Soup for the Soul* had been provided by the publisher at no charge. That didn't appeal to me, and I certainly couldn't go back to what I had taught my sophomores the morning of April 20, war poetry like "Death of a Ball Turret Gunner." Someone mentioned *Jonathan Livingston Seagull*, and someone else eliminated it because it was about two outcasts. That was exactly why I decided it was perfect. It was about two bright outcasts who could have been angry and vengeful but decided to share themselves and offer their gifts to their community.

Of course, I had to factor in PTSD—the kids' and my own. I was barely functioning. I couldn't even handle going to the grocery store. It was too overwhelming and caused me too much anxiety. I wasn't eating or sleeping (unless I took sleeping pills, which left me feeling under water the better part of the next day). Then there was the whole two-second attention span problem. How could I possibly plan lessons and grade papers? How could the kids read?

I decided not to make any of us do either of those things. I could think of no greater comfort activity than a read-aloud. I would read the book out loud every day, and we'd stop and discuss. That way the kids didn't have to try to make their brains follow all that writing on the page, I wouldn't have to come up with involved lesson plans each day, and it would feel like school. I ordered two class sets because I was using them in two different classrooms and didn't feel like lugging them around.

I didn't forget the advice my letter-writer gave, either. My students and I would talk, be honest, try to respect where everyone was. A mental health worker was available in every classroom that first day back, but to my complete surprise, I didn't need mine. The kids walked in, and despite their still-shocked faces, I knew them and they knew me. They were my students and I was their teacher. I hugged each kid as he or she came in, and when the bell rang I said,

OK, guys, we're all in totally different places, and it's going to be that way every day, maybe every few minutes. Some of you got out right away. Some of you were stuck for hours. Some of you saw nothing. Some of you saw far too much. You might walk in feeling sad and suddenly find that you're angry. The person sitting next to you who seemed just fine may fall apart. *You* may think you're just fine and then fall apart. It's OK. I promise to be patient. But I need you to know something. I lost kids I love, and I'm really sad. I'm having a really hard time. I need your patience, too.

They were totally on board and offered their sympathy and support. We were in this *together*.

Then I asked how they were feeling. I heard a lot of "I'm glad to be back at school." I heard "I'm sick of talking about it" and "I can't talk about anything else." I told the kids that both positions were valid. I told them to feel free to write anytime they wanted. If they still needed to process, they could do it that way. If they didn't want to hear about it anymore and someone brought it up, I gave them permission to tune out, draw pictures, do whatever. I reminded them that counselors were available, and if they needed to leave class and see one, they just had to let me know so I could flag someone down to escort them to the office.

Every day, I started class by taking the temperature: "On a scale of 1 to 10, 1 being really upset, 5 being numb, 10 being great, how are you?" Kids could volunteer to give their temperature or not. Those who were really struggling on a given day could make sure everyone knew to be extra patient; those who were feeling good could celebrate while it lasted. It took 5–10 minutes of time, and then I started reading aloud. When they could, the kids followed along in their books, but it was OK to just listen or write or draw. The vast majority did their best to follow, either by reading or listening. Some days the discussion was purely about the story on its most literal level. Others, it was very much about us, about Eric and Dylan, about feeling like an outsider, even among friends, about how hard it was growing up, even without all the things we were going through.

We watched a video of the 1973 movie made from the book, which the kids found hilariously dated. It was good to laugh. The final assignment was around the theme of "my perfect self." It could be a poem or essay, whatever they wanted. One mother, a woman I had gotten to know because I so enjoyed teaching her son, said with the best intentions that she thought I should assign an essay as rigorous as my previous assignments, complete with brutal mechanics grade and grueling corrections. I thanked her for the advice, but I knew full well that in my current state, I couldn't possibly grade essays with my usual thoroughness, and the kids couldn't possibly meet my usual standards. Not gonna happen.

I collected an interesting mix, everything from being a superhero to a more practical hero (like a paramedic) to just being a kinder person. Mixed into those were the classic American dreams of driving a great car and having a terrific family of their own. They could write within the context of the shootings or not.

This, I think, was the gift of my discipline. The study of literature is flexible. A book can be studied in isolation, a work unto itself, or applied to one's personal experience. It can give us a lens through which to view the whole world. Kids could do what they *needed* to with it. Writing assignments can be just as flexible, and I took full advantage of that. I think it's harder to do school and trauma in math and science classes. On one hand, there is something comforting about math—right and wrong are so clear cut. It's a wonderful break from emotion, but it was really hard to introduce new concepts to traumatized brains. Many math teachers simply spent the last few weeks of the year in review; the kids felt safe there.

Foreign language teachers found their own medium: In some classes the kids worked on a travel brochure. It allowed escape; instead of being here and now, kids could think about being in Paris or Barcelona. They were using lots of gross motor skills—cutting and pasting—which are comforting. It felt like kindergarten, like school but a little warm and fuzzy, too, all while keeping it related to curriculum.

In the midst of this, PTSD symptoms were ubiquitous. One of the classrooms I taught in was next door to a room with a spring-loaded door. For the first week, every time someone came or went and the door slammed, half my class dove under their tables, and that was the end of class for the period. The rest of the time was spent trying to make each other laugh to calm the shaking hands and racing hearts. Soon, the kids in the room with the troublesome door learned to hold it carefully while it closed. We let smokers smoke on school grounds under adult supervision. This was no time to get picky about tobacco laws.

In the teacher workrooms, we worked, or stared into space, or cried, or talked. One day, when a group of NFL players just decided to show up in the commons to cheer kids up, a colleague of mine looked at me and said, "If a kid had written this as a short story, I would have told him it was preposterous."

We finished *Jonathan Livingston Seagull* with maybe a week to spare before the end of the school year. By now, I was hearing a strong recurring theme. The kids were so angry with the media they could have spit on every journalist they saw. They were mad at being turned into *bully, jock,* and victimized *nerd* caricatures. They were pissed at having cameras shoved in their tear-stained faces. They were furious at things being taken out of context and showing up on the news as the kind of half-truths that are the harshest of all lies. My final assignment was a critique of five news articles about the event and a summary of what they'd learned about the media in the United States. It gave them a place to vent their vitriol and a common enemy that kept them bonded. It worked. We made it through the year.

It was harder by far to make it through a summer without each other.

## ONGOING TURMOIL AND TRAUMA

The next year was all about keeping our heads above water. Fate doesn't grant special dispensation for trauma served. My father died of pancreatic cancer the next October. That same month, one of my students whose sister had been paralyzed in the shooting lost his mother to suicide. In February, another student lost his best friend to a shooting at the sandwich shop just up the street from the school, all this in a town where violence was a complete anomaly. In the spring, Greg, a popular student, hanged himself. We were just reeling from one event to the next.

Selecting literature for my classes became a challenge. I discovered that *Adventures of Huckleberry Finn* is actually a really violent book, between the shooting of Boggs and the feud between the Grangerfords and Shepherdsons. I hadn't really thought about it before. We cut *Lord of the Flies* from the freshman curriculum for the next three years. The afternoon we learned of Greg's suicide just happened to coincide with my having assigned for that night the chapter in *Catcher in the Rye* where a boy jumps to his death wearing the protagonist's shirt. What can I say? It was brutal, but these literary events allowed for the same thing as *Jonathan Livingston Seagull*. Sometimes we talked about the books. Less often we talked about us, but we could if we needed to.

Teachers were lenient. Extensions were granted, assignments excused. Ultimately, we differentiated for trauma as one differentiates for any disability. Most teachers know real effort when they see it, and our job is to squeeze out just a little more. If we squeezed and the kid seemed paralyzed, we backed off. We often wondered whether we were letting kids milk the situation, but most of us preferred to err on the side of compassion.

Ultimately, no matter what events transpire in students' lives, the school is responsible for making sure they have certain skills and levels of knowledge. Those remain the baseline, but the number of times or ways they demonstrate the skill can be streamlined, if need be. My sophomores had to be able to write a five-paragraph essay. I assigned two of them to hone the skill, but when my student's mother killed herself, I told him to feel free to let the second essay go. It wouldn't hurt his grade. As it turned out, he wanted to write it and did, but it was important to take the pressure off.

Some of this came back to bite us as years passed. Kids got used to a lot of flexibility, even incoming classes of kids who hadn't been there that day. Reining them back in two and three years later was an ordeal in itself. Parents demanded ever-decreasing levels of accountability for their kids, lashing out at teachers for "not understanding." A parent scolded one of my colleagues, informing him indignantly that her daughter had been diagnosed with PTSD. "So was I," he replied. She was shocked speechless. *Teachers* had been diagnosed with PTSD? She couldn't believe it.

Shocked? I was 36 years old. Nothing in my entire life had prepared me for that kind of trauma. Nothing had prepared me to get kids through it. I was *drowning* in trauma, trying to teach and be a full-time grief counselor to over a hundred kids every day, and when I went home, I had to be a wife and a mother. I was a typical Columbine teacher.

## TRAUMA IS PERSONAL

This is as good a place as any to switch gears, bring it tighter to a more personal level. It's a bit graphic at first. I apologize for that. My experience of trauma is this: Being completely unable to track a conversation. Feeling my skin crawl, just crawl, to the point that I wanted nothing more than to rip it off and run away. When I came home from work while we were at Chatfield, I often ignored my husband and two kids (one eight years old and one four at the time), walked upstairs to the bedroom, sat on the floor, and stared at the bedroom window. I fantasized about shoving my hands and arms right through the glass and grating all that irritating flesh off, severing an artery, and calling it a day, a life. I never told anyone that before writing it here because I knew it was crazy. *I* was crazy. That crawling sensation is still my first warning that a PTSD episode is imminent. Trauma is depression that makes every effort Herculean, whether it's grading essays or sorting laundry. It is a wall between you and anyone who didn't go through the event with you. It is a constant sense of suffocation.

I am not a naturally depressed person. I am a caregiver and endless optimist. As a teacher, I am dedicated and competent. After the shootings, I didn't know who this strange, depressed, edgy woman was, and neither did my husband. He was absolutely supportive, but utterly bewildered. I don't really remember much about my kids at all that year. Thank God they both say they don't remember much either. I don't think they'd have very complimentary things to say about me if they did.

I went into therapy in the fall of '99. My therapist asked me over and over about April 20. That wasn't what I needed. I needed to know how to survive my father's impending death when I had already completely ODed on grief. I needed to figure out how to make it through the grocery store. I needed to figure out what to do with emotions that were in the full-on or full-off setting with nothing in between. I desperately wanted my old competence back. The therapy seemed to do more harm than good, but I did get one valuable piece of advice I would pass along to anyone who, God forbid, might find themselves in this spot:

> The therapist asked me, "Do you give kids extra time to do assignments?"
> "Yes," I replied.

"Do you let them skip some?"

"Sure," I said. "They're traumatized."

"Do you give them a hard time if they miss a few days?"

"Of course not!"

"Give yourself the same courtesy," he told me. "You've been traumatized, too."

If it took two or three weeks to grade 90 essays, instead of my usual week, I didn't sweat it. I took a long weekend off and went to a spa resort with my husband. I still pushed myself hard, but I stopped beating myself up. I think that's what saved me.

In the spring of 2000, I took my son on a trip to the Canyonlands National Park in Utah. A colleague was doing the same with his son, and we spent a couple of days hiking together. At one point, I overheard the two boys talking about their fear of having us go back to school and of how it had changed us as their parents. I realized that it was the first time my son could talk to another kid who'd shared his experience as the child of a Columbine teacher. Of all the support groups that were formed in response to the trauma, no one had thought to pull together teachers' spouses who were dealing with their radically altered husbands and wives or their children who were bravely letting their moms and dads go back into a building where one teacher had lost his life.

The second year, I muddled through, largely by taking up a new vocation. On a whim, back before my kids were born, I'd started writing a romance novel. My husband got the brilliant idea that, in terms of therapy, I needed escape more than I needed to relive the horror. He bought me a laptop and urged me to finish the book. Those hours of escape gave me the wherewithal to do my job every day. I finished that book during the second year out and wrote another in the summer of 2001. The second book became my first published novel.

On the very first day of the third year, before we had even started classes, I knew I had made a terrible mistake coming back. I broke out in hives. Soon after the start of school came September 11, and the PTSD returned with a vengeance. The hives became a daily occurrence, almost as soon as I walked into the building. My hair was falling out, and I lapsed into a headache that lasted a full month without relief. My doctor encouraged me to try an antidepressant, but I refused. Why should I drug myself for an event-related issue? Shouldn't I just be able to tell myself it was in the past and get a grip? Anxiety, depression, and suicidal thoughts came back to haunt me. I knew if I went back another year, I was done for, so I took a leave of absence.

## GAINING PERSPECTIVE

For two years, I wrote romance novels, found an agent, and had three books published. I threw myself back into being a mom, watching *Stargate* with my son in the afternoons, playing with my daughter. I cooked real dinners, which required serious shopping, and got over my "grocery store anxiety." I grieved for my father at last. When my leave was up, I eased back in by job-sharing for two years before resuming my job full-time. I also kept writing and have had another book published. Straight historical fiction this time. I let go of the "happily-ever-after" crutch that had gotten me through the worst of it.

There was one last event that has caused me to seriously alter my view of using medications to deal with trauma. Toward the end of the third year, right before I took that leave, someone came to the school and explained in careful detail with MRIs the neurological causes of PTSD. They showed us how a traumatic event actually reroutes the neural pathways of the brain. We learned what a PTSD trigger does, how it lights things up, stimulates the fight-or-flight response that makes my skin crawl and makes me want to run until I drop. It was helpful, made me feel less like a freak. I didn't fully comprehend it, though, until several years later.

We had a bomb threat at Columbine in 2007. Actually, we got those a lot, but this one was serious enough to evacuate the building. The entire school tramped off to the designated spot—that nearby elementary school that had become the gathering place for kids and parents back on April 20, 1999. The day of the bomb threat, that school was the predesignated gathering place in case of evacuation. The process was much more organized than things had been eight years earlier, but the event was chock-full of my own personal triggers. I was back to crowds of kids and anxious parents. I was back at the same elementary school where one of my dearest students told me Rachel was dead and I called him a liar. By the end of the day, I was back in the same gymnasium where a man had told mothers and fathers to get their children's dental records.

I fell to pieces, wept and shook uncontrollably, and nothing—*nothing*—could move me to get myself together. The next day, during the follow-up faculty meeting at Columbine, it started again. No matter what I did, I couldn't get a grip: skin crawling, hands shaking, tears pouring nonstop. I gave up and called my doctor, who gave me a prescription. I almost never take it, but when I feel myself losing my grip, it makes a world of difference. It doesn't make me high or drowsy or any of the things I thought an anti-anxiety medication would make me feel, but it helps me feel like myself and gain perspective so I can go about the rest of my life. I can't help but assume that I'd have suffered fewer hives, lost less hair, and had fewer headaches back in 2001 and 2002 if I'd taken the medication then. I really think I could have circumvented the physiological issues and been better equipped to deal with the emotional ones. My final piece of advice to trauma victims: If your doctor prescribes medication, take it.

Since then, one of the ways many of us teachers deal with what happened is to share it with those who need it. When a girl was shot and killed at a Colorado school by a stranger who had slipped into the crowded hallways undetected, we arrived with lunch and comforting words that very few could honestly say to them: We know what you're going through. When a student went on a shooting rampage at Virginia Tech, I contacted the English department chair. Later she told me that my perspective, warning her about how difficult it could be to integrate a new freshman class into a school dealing with such an event, really helped. The school took a number of extra steps to smooth out that integration.

I wish I didn't know so much about making school happen after a shooting, but as Gerda Weissman Klein, a Holocaust survivor who kind of adopted Columbine, often told us, pain should never be wasted.

\* \* \*

### CONTRIBUTOR NOTES

**Paula Reed** teaches English, competitive speech, and debate at Columbine High School in unincorporated Jefferson County, Colorado. She has authored three romance novels and the acclaimed novel, *Hester, the Missing Years of* The Scarlet Letter.

### WHAT'S NEXT?

Section Two offers an opportunity to learn from the stories of reclaiming school after large-scale, community-wide catastrophes as well as individual experience with trauma. It begins with an account by Jethro K. Lieberman, administrator and professor at the New York Law School, not far from the World Trade Center, who assisted in restoring services and responding to the needs of students, faculty, and staff after the September 11, 2001, terrorist attacks.

# LEARNING FROM TRAUMA

CHAPTER 4

# NEW YORK LAW SCHOOL AND THE CRISIS OF SEPTEMBER 11, 2001

## JETHRO K. LIEBERMAN[*]

### SNAPSHOT OF THE CRISIS

New York Law School, an independent law school founded in 1891, stands in the Tribeca section of Manhattan about eight blocks from the site of the World Trade Center. On a mild, sunny Tuesday, September 11, 2001, students and a few faculty members and staff who had arrived early to get coffee before classes had an unobstructed view of the incomprehensible: a jetliner slamming full speed into the upper reaches of the north tower at 8:46 a.m., followed less than ten minutes later by a second suicide run at the south tower. Shortly after 9:00 a.m., our main switchboard was alerting callers with a message saying that the Law School was closed—the first non-weather-related term-time closing in our history.

It was nothing we had ever anticipated—or planned for.

That morning I was ambling along in my car a little before 9:00 a.m. and, as is my habit, listening to the traffic news on radio station WINS. The announcer suddenly spoke of a mysterious accident that had just befallen one of the WTC buildings. I wasn't really concentrating, and it didn't seem relevant, since the buildings are not on my route—at least not on my driving route. But they loomed large at the end of my subway ride, a half a mile straight south of the subway exit. A couple of minutes later a more agitated

---

[*]Author's Note: A few portions of this chapter are reprinted from the volume *Eight Blocks Away*, with permission of Tribeca Square Press and New York Law School.

voice announced that an airplane had crashed into the WTC. That was odd—very odd—but still, nothing that seemed a portent of the horror to follow. A few more minutes south along my way to the Bronx, where I pick up the No. 1 train, which runs straight down the west side to my office, we listeners were given a breathless account of a second airplane hitting the second tower. Still I drove on, reaching my parking spot at 242nd Street near the subway station at about 9:10 a.m.

Now convinced that something serious had happened, I tried to call the office on my cell phone. I could not get a circuit. I found a payphone, which to my surprise gave me a dial tone. (Payphones in Manhattan in those days were on their last legs and have now all but disappeared.) I called my assistant. A message came on, reminding me that she had planned to be out that day. "For a live person," her voice said, "press the star key." When I did, a recording came on, saying that the Law School was closed. It was about 9:15 a.m. It was official. Something horrible had happened. Until that day, the Law School had never closed for anything other than snow, and large amounts of it at that.

Back in my car and headed north, I learned that the Pentagon, too, had been hit, and for the first time I heard the words, "terrorist attacks." Home about 20 minutes later, I tried to phone my wife at her office in upper Manhattan, but the circuits were jammed and remained so until the next day. I finally managed to send her an e-mail message that I was okay—the Internet worked. Otherwise, I was cut off. The TV showed me the collapse of the towers. After 15 minutes, I turned it off. I figured I'd learned all I was going to learn until much later in the day.

What we learned in the days—and even weeks, months, and in some cases years—that followed was that we had entered a new era in human history. Al-Qaeda's 19 suicide "soldiers" in four hijacked airplanes killed nearly 3,000 people that day: More than 2,600 were killed in the WTC towers and on the surrounding streets, about 125 at the Pentagon, and about 250 in the planes (two rammed the towers, one smashed into the Pentagon, and one was diverted and crashed in southwestern Pennsylvania near Shanksville). Of the people killed in the buildings and in the rescue efforts in New York, 411 were emergency workers, 341 of whom were New York City firefighters.

The suicide missions sparked a "war on terror," a long chain of events that led the United States into two prolonged wars, in Afghanistan and Iraq, enactment of the Patriot Act that critics charge has seriously eroded constitutional liberties, caused a serious decline in American moral standing around the globe, and created conspicuous difficulties in air travel— the most obvious among the myriad of post-attack developments. But on September 11, 2001, all that was in the future.

If you put aside the horror and the chaos of the hours that followed on the day itself, the eeriest aspect of the attack for those of us in the neighborhood, or so it seemed to me, was the utter silence and stillness that followed. By late afternoon, in a spooky delayed reaction, the networks that connected the Law School to the outside world collapsed, though not all at once. Electricity went down; computers were so many hunks of metal and mute wires; information about students in our administrative systems became inaccessible; the Internet connection was severed; our website vanished; phone lines were dead. In the immediate aftermath, we could neither stay at school nor come back to school even if we wanted to.

The following Monday, September 17, when many of us did return for the first time, the closest we could get by public transportation was the subway, which stopped at Canal Street, about four blocks north of school. As we came up the steps to street level, we were met by police barricades and a phalanx of blue-clad officers. They were very polite and very stern. They would let you through if you showed them two forms of identification. One card had to have a picture and the other had to show that you worked within what we now understood was a war zone.

The roads, which less than a week before had been choked with traffic, were empty, the sidewalks deserted. The city had imposed a no-drive zone below 14th Street, about a mile north, and it only gradually shrank over the succeeding weeks. Scenes from post-apocalyptic movies played in my mind. You looked straight down abandoned streets, a ghost town that less than a week before teemed with life. The characteristic Manhattan decibel level, the rumble and rush and screech and babble of vehicles and pedestrians that make the city what it is, had been silenced, as if someone had thrown a mute switch across a suddenly barren land.

In human terms, we were collectively lucky. No student and no member of the faculty or staff died or was seriously injured, though we did not know that immediately. We later discovered that four of our alumni, working in the World Trade Center towers, had been killed that day; and, of course, we all knew someone, or knew someone who knew someone, who had died.

In property terms, too, we were lucky. Our buildings were intact; structure, equipment, furniture, and supplies all undamaged.

But in business terms, it was far from clear that luck was with us. The buildings, though standing, were inert. Without power or communications, what good were they? And even if they could offer a place for education, how would anyone get to them? Those who depended on automobiles would be forced to abandon their cars far from our doors. Train lines, especially from New Jersey, were seriously disrupted. Subways stopped without warning. Already lengthy commutes could consume significant portions of a day. How were we going to get students, faculty, and staff through the barricades and back into classrooms? How would they catch up with the relentless pace of a law school schedule? Students could not access their lockers, could not read the books they had left behind, and even those who were paying for Internet service at home had nothing in those days like today's access to Google searches and downloadable PDFs from publishers or even from friends across town. Even more elemental, how were we going to let our students know that classes were resuming? And if we managed just that, would they come?

Reminders of the horror were everywhere. The police were omnipresent. Roadblocks and barricades built a new urban landscape. In the immediate days after, ash lay on the streets. Even more than the awful stench that persisted, though weakening in intensity, for three or four months, fear was in the air. We were safe for now, but for how long would our luck hold out?

At a meeting of some of the faculty and senior staff on Thursday, September 20, four days before classes were to resume, a long-time employee stood up and said she was afraid to come back, "now that we know how insecure we are and how easily we can be attacked" (or words to that effect). I sought to allay her fear by observing that we actually were much more secure. We were insecure *before* the attack—we just didn't know then that we were vulnerable. But after all that had happened, it seemed highly unlikely, I surmised, that these villains, or any other villains, could mess with New York City. We were now protected by an alert and vigilant police and military presence. I doubt she was reassured, though for a decade, thanks to just that vigilance and some good luck, my prediction has so far held up. Many others shared her concerns. It didn't help that we had had to abandon our initial meeting three days earlier because of a bomb threat that had been phoned in to ConEdison utility workers attempting to restore full power to the building. The threat turned out to be a false alarm, but what once would have been dismissed out of hand, or taken in stride, now sufficed to keep most people on edge, and for quite a while.

A month later, rumors spread that the FBI had concluded there would be another attack on the city on a particular day in October, and many people stayed away. The rumors were empty and nothing happened. Two months later, many automatically assumed that the crash of American Airlines 587 in Queens on November 12, shortly after taking off from JFK International Airport was a renewed act of terror. Even nearly three years later, in the spring of 2004, when we were contemplating a move to larger facilities across the street from the New York Stock Exchange, some of our colleagues were apprehensive about relocating to what seemed to them a potential al-Qaeda target. In the end, and for other reasons, we did not move further downtown, choosing instead to build a new classroom, library, and events facility on land we owned adjacent to our existing buildings in Tribeca.

Before September 11, the only type of crisis for which we planned and regularly rehearsed—likely the only type of crisis that most schools regularly contemplated—was the prospect of fire. Like everyone else, we held and continue to hold routine fire drills. But as we all know, fire drills are highly specialized. We learn to get people out of buildings that are on fire, to summon help, and, if we are good at it, to account for occupants to ensure that they really are out on the streets and not trapped inside. When the drill concludes after about 30 minutes, we dutifully return inside. We do not practice for conflagration—for the prospect that we will not be able to re-enter and resume our work. But that's what happened, on September 11, 2001.

The day of the attack, it was nearly impossible to reach anyone by phone in the New York area because the mass of calls jammed the circuits. But by the next afternoon, the senior management team—Dean Richard A. Matasar and eight or nine of us who oversaw the Law School's operating departments—held our first conference call, and continued doing so daily for the next week. These calls lasted two to three hours or more, as we sought to construct a plan for renewal, both for the individuals (students, staff, and faculty) who comprised our community and for the institutional systems and schedules.

Our problems were apparent. Though we had a staff and faculty directory on paper with home phone numbers, we did not have an offsite master list of student phone numbers or alternative e-mail addresses. Our website was dark, and our servers were down, so we could neither broadcast news nor reach anyone through the Law School e-mail system. Our phones were also out of service so we could not record any outgoing message on the school's main number. We did not know how long we would be barred from returning to our buildings by order of the mayor or by the lack of power, so we did not know how seriously affected the class schedule would be for the semester. We did not know what our students were thinking or feeling, or what we ought to be doing for them, or how we could do it. We did not know how many students remained in the area. We did not know how many would return when the dust (literally) settled.

New York Law School has had a long history. By 1904, it had become the largest law school in the country (measured by numbers of students). Our founding faculty, a group of disaffected professors from the Columbia University Law School, attracted a number of prominent outsiders as lecturers, among them Woodrow Wilson and Charles Evans Hughes. Several "name partners" of the city's prominent and still leading law firms were graduates, as were such political leaders and jurists as Senator Robert F. Wagner and Supreme Court Justice John Marshall Harlan II. The Law School was one of the earliest to run a full evening division for working students and also one of the earliest law schools to admit women and African-Americans. It catered, as it still does, to immigrant students who were the first in their families to study beyond high school. But the World War II draft created difficulties that forced the Law School to shut its doors for the duration. When it reopened in 1947, it faced a long climb back to the ranks of an intellectually vibrant establishment of higher learning. At the start of the 2001–2002 academic year, it was a school of 1,402 full-time and part-time students, 59 full-time faculty members, more than 84 adjunct faculty, and 128 administrative and support staff, and one very real question was whether the terrorist attack would knock us out once again.

Whatever most of us could do, it would be at a distance, and it would be very much by the seat of our pants. Our challenges were many, and for the sake of clarity, the account that follows is grouped by topic. In reality, of course, we dealt with all the problems simultaneously.

## DEALING WITH THE CRISIS

### COMMUNICATIONS

We needed to find people. Some connections happened naturally, normally, as staff, faculty, and students reached out to whomever they could find. By the next day, we were talking to some staff and faculty, and faculty talked to some students or heard from them through alternate e-mail addresses. But there was little of concrete value that we could say in the immediate aftermath. Before the Law School's servers went down on the afternoon of September 11 (at 3:44:47 p.m., as the message header recorded it), Dean Matasar sent an e-mail to the entire school community: "As I look out of my office window I can see the smoke and debris from the Trade Towers. There is no worse feeling than the helplessness that comes from being close, but unable to help." He acknowledged the grief that was upon us and expressed the hope that "the rule of law be followed," that our leaders would "respond swiftly and in a measured way," and that the "rest of us—New Yorkers, Americans, and all members of the legal profession—use every tool at our disposal to help bring our community together to respond to a shared national catastrophe." But there could be no actual plan at that hour. The most that could be hoped for—as people were still fleeing the area and the city—was that "all of us in this community work together and come back as strongly as we can."

By late the next afternoon, after the deans' first conference call, as associate dean for academic affairs, I was writing to the faculty, not knowing how many would be receiving the e-mail, a collection of sketchy plans—actually more hopes than plans. Perhaps we would resume classes a week hence. The schedule would need to be dealt with globally; faculty were urged to refrain from attempting to schedule makeup classes individually. The New York Court of Appeals and the American Bar Association's law school accrediting arm take seriously our adherence to a minimum number of class hours, and not even this emergency would necessarily allow us to cancel some unknown number of class meetings. Perhaps providentially, the existing schedule already provided that there would be no classes the following Monday and Tuesday because of Rosh Hashanah, a holiday observed by school closings in New York City and the region. As it turned out, we missed eight days of classes, few enough that we could and did manage to make them up by using weekends and extending class hours throughout the term.

I was also able to let people know that an alternative website would go *live* shortly. It was being provided by the company through which we out-sourced our informational technology (IT) services and therefore would not immediately be accessible through the school's URL. That would help but not entirely, since we would still face the task of notifying everyone to look for the new address. Luckily, as it turned out (I spare the technical details), our IT director had, by chance, about three weeks earlier made changes to our "ISP profile" that permitted it to redirect our website from our primary server at the school to our outsourced IT company's servers in Florida, where a complete copy of our website was maintained.

On Thursday, we learned that because our upper-class students had all been trained on the Westlaw online research system the previous two years, the company had those students' home e-mail addresses. (Westlaw is one of the leading providers of legal databases to the profession, including all law schools.) This was big help, since we could now create a distribution list for two-thirds of our students. But the first-year students, who had been on campus less than three full weeks, had not yet begun that training, and we still did not have *all* their addresses. (Some teachers of first-year courses did have e-mail addresses for their students.)

By Friday, September 14, our website was switched from its temporary address back to the school's URL familiar to our students, staff, and faculty, though we could only access it through our outside IT company. On that day, also, the telephone lines were restored, and we posted a message to all callers to check the website for updated information. The website for the first time provided the URL of the alternative website, and we asked students and others to jot down that address just in case the current site failed. In our daily conference call, we concluded that we could not sensibly plan to schedule classes to begin the following week, so we postponed the restart until Monday, September 24.

## FACILITIES

Several staff members, including the director of facilities and maintenance, the IT director, computer technicians, the director of telecommunications, and the associate dean for public affairs, gained access to the buildings on Thursday the 13th. One of the principal objectives was to restore power to at least some portions of the space so we could access our databases. To that end, a heavy-duty, trailer-type generator was brought to the sidewalk, and after hours of effort, power was partially restored, giving us hope that we could soon host the school website through our own servers. Fred DeJohn, associate dean for finance and administration, reported the next morning: "Of course you knew this would happen! At midnight last night I saw on TV that the streetlights in our area had suddenly come back on. I immediately called School and learned that the power had just come back to our building." [Only about eight hours after the generator had finally begun to work.]

Nevertheless, the generators were kept on through the weekend "because we need to get the electricians back in to reverse everything that was done yesterday." Unfortunately, the T-1 Internet lines were still not operational. They were routed through a relay station that had knocked out phone service to some of our buildings, including the server area, and that was still not functioning properly. Our alternative websites would have to serve for the immediate future. It took weeks before the Internet, telephone, and fax connections were fully operational.

## AIR QUALITY

The good news was that a structural engineer's check of the buildings found nothing amiss. Once power was fully restored, everything would work as it had. What remained unclear was air quality. Various official statements, on which we had no reason not to rely, proclaimed that the air where we were was safe to breathe and that only those at the immediate World Trade Center site were supposed to wear masks. But the odor was pronounced, and for weeks afterward several members of the school community sought to have air purifiers installed throughout the buildings. The odor lingered for three or four months, but it was not until some years later that we learned that the environmental "experts," and the officials who gave their opinions, including the then chief of the U.S. Environmental Protection Agency, were either uninformed or, worse, less than truthful.

## REASSURING STUDENTS AND OTHERS

None of what we were doing in those first few days would make any difference if students, staff, and faculty were too upset or frightened to return. For the students in particular, the attacks were shattering, deeply wounding their essential sense of the United States and the world around them.

Most of them had been born in the late 1970s and had come to social and political awareness in the 1990s. They were a generation of Americans for whom the country had never been other than invincible and prosperous. *This could not be happening* seemed to be the common thought. *This just isn't fair. There were things to be done. ... This wasn't part of the plan.* I quote from a long e-mail that arrived on Saturday from a shattered student to one of the associate deans:

> Hello, I am a student in his first year at NYLS. I am very concerned and afraid of going back to class at NYLS as early as the twenty-fourth. The horror and panic that I experienced that frightful morning on my way to school will never be forgotten. ... I really don't know what to think about school. I have no books anymore and just feel like these terrorists have taken so much that I love away. I am afraid they may strike us again. ... Can we possibly focus on normalcy while our walk to the building includes passing the war zone all around us? I hope I can make it, I hope we can all make it, but there must be an extremely high caliber of assistance from our School Community. ... This is going to be hard. I know it is, we cannot just be expected to dust off, place our families' concerns aside and at the same time sustain the pressure of law students successfully. You all know that we are all going to be a mess, there isn't a day I am "normal" lately. Our building still stands but our hearts as students are at ground zero. Be conscious of this above all. Please write me back. I haven't heard from anyone.

We realized from our first conference call that we could not simply announce a day on which we would reopen and expect students to walk into classrooms as if they were returning from an extended break. Several faculty and staff members urged a number of initiatives designed to allow students to confront what had happened. First on the list was to arrange for a number of mental health professionals to be on hand for students and employees once classes resumed. Many such professionals volunteered their services and were available in one-on-one counseling sessions for several weeks. However, we later discovered that many students were reluctant to avail themselves of mental health counseling because at that time applicants for admission to the bar in New York State were required to declare whether they had ever sought such assistance, and the students were fearful that answering *yes* would somehow jeopardize their ultimate admission to the bar. (This question has since been removed from the application.)

On the Friday before our reopening, we held a "teach-in" so that students could air their concerns, fears, hopes, and other feelings. It was widely attended, and other large group meetings followed. Every professor was encouraged, likewise, to welcome students back by reflecting on the attack, to let students talk through their feelings and reactions and, as appropriate, to think through how the law and the legal system could deal with terrorism and could help bind the wounds that had been inflicted. We also planned a memorial volume to be published a year hence, and invited every member of the community for whom writing about what they saw and thought and experienced would be helpful to do so. That volume, *Eight Blocks Away*, was privately published a year later, and was reissued, by Tribeca Square Press, in time for the tenth anniversary. Sixty students, staffers, and professors contributed. (I have relied heavily on it for this account.)

No less important, we sought to provide avenues for students to act on their desire to be helpful, a common but nevertheless ennobling impulse felt by many members of communities struck by tragedy. In particular, on the Friday before we reopened, Professor Stephen J. Ellmann, associate dean for faculty development, convened and chaired in our newly stirring buildings, the first meeting of an organization created for that purpose: the September 11 Law School Pro Bono Coordinating Committee. In attendance were 40 people from eight bar and public interest groups and ten law schools, who came together "to discuss the legal needs created by the World Trade Center attacks and the contributions that law schools could make to meeting those needs."

Some projects were already underway, staffed and funded by various organizations in attendance—among them, a volunteer legal assistance program established by the New York City Bar Association. Most of the early calls were for landlord-tenant matters and for helping survivors with crime victim assistance forms and expediting issuance of death certificates. Many other projects were discussed, such as facilitating "legal inventories of the needs of individuals and families harmed by the attacks"; assistance for small businesses; creation of hotlines for people seeking volunteer attorney referrals; assistance in allocating disaster funds "to support provision of legal services to poor people"; support for the needs of undocumented people and the survivors of undocumented victims, including immigration problems that would result from the death of a citizen spouse; help for those who lost employment in the attack; assistance for the newly homeless in matters relating to mortgages and rent; help with a range of insurance issues, including workers' compensation, Social Security claims, and health care; advice and assistance on a wide range of other matters, including food stamps, welfare benefits, tax issues, probate of wills and of estates left behind in the absence of wills, foster-care problems if the parent in a single-parent family was killed, and much more.

It was a daunting list, and the Coordinating Committee pledged to work with each other and to place students, among others, in a host of volunteer efforts to tackle all these and other concerns. The National Association for Public Interest Law in Washington, DC, honored New York Law School and

Pace University Law School, which sponsored similar efforts, with its 2001 PSLawNet Pro Bono Publico Award, for undertaking these programs. On Monday, September 24, classes resumed. The schedule was altered a bit to permit make-ups for most classes missed (it was not possible to do so for every class), and to extend deadlines for various writing assignments and other projects that would have been due during our hiatus. Early on their return, the faculty took up the question of what to do about the emotional effect of the attack on the students' ability to regroup and return to their studies. The result was a resolution to suspend normal academic rules for the semester and to permit students to be graded in one of their courses on a pass-fail basis, with freedom to select the particular course after their final grades were posted, thus permitting them to avoid the effect of the lowest grade on their GPAs. Students expressed gratitude not merely that this policy was adopted but also that the faculty took their concerns seriously.

As one second-year student wrote the dean in an unsolicited letter six weeks after,

> Through it all, I looked to you and the staff as beacons of strength and normalcy. You and your staff provided information, comfort, and a sense of safety while just outside our door the world was coming to an end. After school reopened, I was again impressed at the school's response. It was wonderful to see the administration and staff take such strong interest in the students' physical and mental well being. Having counselors available upon our return was and continues to be a great help. Equally as wonderful have been the teach-ins, workshops, and all the opportunities to discuss these horrific events, and hear other people's perspectives. My mother came with me to the Friday teach-in just before school started, and both my parents accompanied me to an October 3rd symposium at the school. They too have been impressed and thankful at the opportunities provided.

It all seemed to work, though it would be weeks before the phone lines and Internet service were fully restored, months before the air smelled normal again, and even years before the final police barricades came down in the blocks surrounding our buildings. Everyone rallied, and by the spring semester it was possible to pass an hour or two—perhaps even more—without thinking about the tragedy eight blocks to our south. Almost the entire student body came back and stayed to graduate at their appointed times. Our records suggest that only ten students failed to return, deterred by the horror and angst that befell us on September 11, 2001. We remain in a much expanded campus in Tribeca, our square footage having more than doubled with the completion in 2009 of a new building and the taking of a long-term lease for an entire city-block-long floor in a building across the street.

For many of us who were there on that day, the memory remains fresh. For students who newly enroll, the day is receding into history.

## LESSONS LEARNED: THINKING THE UNTHINKABLE

Some time later that academic year, all the deans got BlackBerries, a device that would permit us to read e-mail portably. We had resolved that our communications would not be at the mercy of future shut-downs, though in the Great Northeast Blackout of 2003 it became evident that nothing is perfect. When the entire eastern seaboard power grid went black, so did the means of directing calls and e-mail to cell phones. Our first BlackBerries doubled as walkie-talkies, but the dean relented when it quickly became clear that most of us were all thumbs and balked at saying such things as "over" and "out." We quickly replaced them with new BlackBerries, minus the walkie-talkie.

More sensibly and more long term, what really did change was the sophistication of our general communications. Our networked files, including administrative files with all student information, are backed up daily and stored offsite. Our website is hosted in multiple places on campus, and it and our mission-critical administrative systems could be rebuilt in about a day if all our buildings were flattened. Around the time that we observed the tenth anniversary of 9/11, our website was hosted at another remote location as well. But no sooner do these modes of communication become perfected than they become passé: If an emergency happened tomorrow, says our director of information technology, we would likely be meeting in the cloud or on social media sites like Facebook and Twitter. Separately, through a system called ConnectEd, hosted offsite, we can post messages to the entire community's e-mail, text messaging, home, and mobile phones within minutes of any disaster. If there is one permanent legacy of the crisis of September 11 for New York Law School, it is a communications system that can survive the physical destruction of our campus.

The other legacy, whether permanent or not remains to be seen, is the knowledge that although we can try to anticipate contingencies, there can be no ultimate, perfect game plan for a crisis that strikes without warning. Since the day the World Trade Center was destroyed, we have faced at least four logistical difficulties, but none that amounted to a true crisis, since each could be anticipated. The first was a threatened strike by subway workers in 2002 against the New York City transit system. Many of us spent hours thinking through contingency plans, but in the end the strike never materialized. Second, in 2004, the city hosted the Republican National Convention, and word of protests and the likely police response to the protests—some of it converging on major subway lines to the Law

School, again with the threat of stoppages—led us to cancel classes on a couple of days, to the dismay of some, in an attempt to avert what might have been serious gridlock and largely empty classrooms. Third, in 2005, the transit workers actually did go out on strike. The shutdown began on December 22, at the tail end of our exam period and during the busiest pre-Christmas shopping week. Our earlier thinking informed our response to the transit shutdown, which did indeed provoke huge delays and gridlock in large parts of the city. Ahead of the strike we postponed certain exams, rescheduling them for after the New Year. On that one, we guessed correctly, though not everyone was pleased with the plans put in place. The fourth logistical problem was the prospect of widespread absences from the potential 2009 swine flu epidemic. Since the disruption would depend entirely on how extensive any epidemic would be, we realized that although we could take certain steps to minimize the spread of flu within our premises, there was nothing we could do beforehand at any sort of reasonable cost that would permit classes and operations to continue unimpeded. We would have to trust to good fortune and, that failing, plan as we went along. Happily, good fortune prevailed.

But these were the easy ones, because in each case we had advance warning of the peril and hence at least some time and capacity to think through our responses. We are even now drafting a multipart emergency operations plan consistent with post-2001 federal, state, and local laws and regulations. It can guide, but it cannot prevent.

The difficult problem, for which there can never be a general solution, is the crisis that strikes without warning or the dimensions of which cannot be foretold. Like a fire, a blackout, an earthquake, a coastal storm, airborne chemicals, or a mad gunman—certainly, physical and human security and emergency systems must be in place, but in the end, to survive the crisis, an institution must trust in the trained judgment of responsible administrators and faculty.

*   *   *

### CONTRIBUTOR NOTES

**Jethro K. Lieberman, J.D., Ph.D.,** is professor of law at New York Law School, where he has been teaching for 27 years, and was associate dean for academic affairs during the 9/11 crisis. He currently is vice president for academic publishing. Before beginning his teaching career at Fordham University Law School, he was legal affairs editor of *Business Week Magazine*, vice president and general counsel of a small New York City-based trade publisher, an associate at a large Washington law firm, and was on active duty in the Navy Judge Advocate General's Corps. His 25th book, *Liberalism Undressed,* is forthcoming from Oxford University Press in 2012.

\*   \*   \*

**WHAT'S NEXT?**

Large-scale disasters, whether caused by humans or the result of natural forces, interrupt normal functioning, sometimes for extended periods of time. Bridget Ramsey's narrative "Watermarks: Leading and Teaching in the Aftermath of Hurricane Katrina," provides a look at the challenges she faced and strategies she employed as the director of a public school that reopened as a charter school after the devastating storm. Meeting the needs of traumatized students, faculty, parents, and community was complicated by inadequate resources, overwhelming poverty, and loss of basic infrastructure throughout the area.

# CHAPTER 5

# WATERMARKS: LEADING AND TEACHING IN THE AFTERMATH OF HURRICANE KATRINA

BRIDGET DWYER RAMSEY

THERE IS NO DENYING THAT HURRICANES ARE A DESTRUCTIVE force of nature. As a child growing up in New Orleans in the 1960s and '70s, I knew summer months and early fall were accompanied by storms churned up as the waters of the Atlantic met the comparably tepid waters of the Gulf of Mexico. Still, at the approach of a storm, my family, like most others during those years, never evacuated the city.

The news of a hurricane approaching often filled us with excitement and anticipation. How windy would it get? Was it really possible for the roof to blow off? Maybe we'd never have to go to school again! For a school kid, the prospect of a storm seemed a marvelous adventure. I recall a rare day or two of early school dismissal, my mother lighting the votive candles on the mantle, my older brother and me helping haul mattresses downstairs to a safer location in the house, the greying of the sky, the waiting, the eventual flickering off of the electricity as the wind and rain pelted the house, the sleepless night, and finally, the calm of the morning that revealed uprooted trees or collapsed sheds and other fixable damage.

As the years progressed, weather forecasters became better equipped in their "pinpoint accuracy" to assess location and severity of approaching storms, and evacuation became more common. When the weather became the news, it was time to pack the car, put some food in the cat's bowl, leave the bathtub filled with water, and go visit the relatives. We'd be back in no time. The hurricane would surely dissipate. All would be well.

But on August 29, 2005, that script was rewritten by the forces of Mother Nature combined with the collapse of inadequately constructed levees. While hurricanes are a way of life along the Gulf Coast, few people ever imagined that a hurricane could silence a city, halt commerce, wipe beach towns off the map, disrupt schooling for five months, and reveal to the world scenes of extreme desperation and hopelessness. Hurricane Katrina, the Category 4 storm that swept over the city during the early morning hours, and the floodwaters that breached the levees in the days following, did just that.

The destruction, disruption, and hardships wrought by the storm are indelibly marked on the collective consciousness of everyone who lives in "the Crescent City." Seven years later, many people continue to sort out their post-Katrina lives; but rebuilding and renewal are also evident, most notably in the realm of public education.

Poverty and a history of educational neglect exacerbated problems encountered in the aftermath. Many areas that were flooded by the levee breach had once housed thousands of economically disadvantaged people. Neighborhoods were destroyed, and entire communities disappeared. The pain and disorientation that students and adults felt were apparent everywhere. A chance to help rebuild educational opportunities for the children of New Orleans is what brought me back to the city. I accepted a job as the director of a post-Katrina "conversion" charter school—New Orleans Science and Mathematics High School.

In the aftermath of Katrina, charter schools represented an expedient way both to reopen schools and to begin providing the city's children with what the governor called "a world-class education." A state takeover of the majority of the city's schools following the storm essentially dissolved the old system, but instead of an orderly transition toward positive reform, the storm created a crisis that, in effect, led to closing the failing schools without a coherent master plan for opening better ones.

### GETTING STARTED

My enthusiasm, dedication, education, and 16 years of experience as a teacher and administrator in public and charter schools did little to prepare me for the instability and awaiting challenges. The hastily written and quickly approved application that had authorized the conversion of my school to charter-status was well-intentioned. The city was experiencing a slow trickle of returning residents, and students needed to be in school. The school district admitted publicly that it lacked the resources, personnel, and brick-and-mortar buildings to open schools for the remainder of the 2005–2006 school year, and perhaps for even a year beyond that. Using federal charter start-up funds to convert traditional schools into charters appeared to be the best option.

Driven and determined educators stepped in to open schools under the most daunting conditions imaginable: flood ravaged buildings standing in filthy water, sodden student and personnel files, ruined textbooks, and moldering computers. The majority of students arriving to register came without any documentation of their academic status, and for some, the only proof of their name was a grandmother or another relative who accompanied them.

Prior to Katrina, the school I would lead had served for 13 years as a half-day school (morning or afternoon classes) for students seeking enhanced education in science and mathematics. While not all students had been gifted, would-be scientists or mathematicians, with the guidance of a stellar cast of educators, students flourished, developing a deep understanding of the subject matter. The school's previous success would stand as the foundation on which it would open and build into a full-day, four-year, open-enrollment, comprehensive high school.

In January 2006, four months after the storm, the school opened its doors to a handful of former students and any others who applied for open enrollment. The plan was to grow the student population to 400 over a period of four years. In an effort to instill a sense of pride, school affiliation, and order, students were required to wear uniforms. Student services kept a closet stocked with uniform clothing for those students in need, but supplies were quickly depleted.

Students were still slowly making their way back into the city in July of 2007 when I took over as director to replace the previous director who was retiring. I met with her numerous times, listening to her vision as the founding principal for the new school, and making plans to transition into the leadership role. We were in tune with the commitment to open enrollment, rigor, remediation to get kids where they needed to be, and the emphasis on math and science as subjects critical to their future. The school had a dedicated foundation with a track record of raising significant additional funds for summer internships for the juniors and seniors in the health and science fields. I introduced the concept of Science, Technology, Engineering, and Mathematics (STEM) education that was and continues to receive serious attention and funding in the realm of school reform and educational excellence for twenty-first century learning.

After I officially became director, I began to take stock. Students continued to show up without transcripts, birth certificates, social security cards, vaccination records, or proof of residency. Many were in the process of searching for a place to live or were living in trailers by themselves or with a relative or a friend, and in New Orleans those terms were greatly stretched.

A survey of students indicated that 52 were homeless, and there were probably more. The definition of *homeless* is having lived in three different locations in the previous year, and the majority of the people in New Orleans were in a transient situation, since livable, affordable housing was scarce. Trailers were provided as temporary housing by the Federal Emergency Management Agency (FEMA). Teachers, too, were living in FEMA trailers, their homes and lives as disrupted as the students'. The school was a microcosm of the city—a small community of traumatized adults trying to establish order and a sheltering learning environment for dismayed youth longing to reconnect with friends and routine.

For the students, trauma from their life experience showed itself in relational aggression, physical posturing, disruption of classes, fighting in the hallways, and lashing out. In speaking with colleagues leading other schools, I learned that they, too, were witnessing student behaviors associated with post-traumatic stress disorder: lethargy, inability to concentrate, thumb-sucking (high school students), excessive nail biting, emotional outbursts, hypervigilance, or paranoia ("That teacher's out to get me"). Yes, there were students—and teachers—who outwardly appeared to be coping, but an undercurrent of uncertainty and an urgency to get back to normal were ever-present.

The governing board hired me with the expectation that I would put a curriculum in place, oversee the training and instructional accountability of teachers, and get adequate science labs for the building (which had been an elementary school). I was to guide the school toward success that would ensure its future as the preeminent science and math school in the city.

I quickly discovered that no one on staff had ever scheduled more than 150 students for classes. Compounding the seriousness of this skill gap was a mandate received from the state-run Recovery School District to take on an additional 125 students. This meant hiring more teachers in a city where effective teachers were in short supply and competition for them was voracious. In an attempt to attract math and science teachers, signing bonuses were offered and salaries raised. In some instances, people with industry backgrounds and no high school teaching experience were hired.

We started the 2007–2008 school year with approximately 300 students and 33 teachers. Without a software program for scheduling, the task had to be done by hand. The lack of transcripts made determining grade levels and assigning students to classes something of a guessing game. This situation contributed to the frenetic pace of the school day. Students might be assigned to a class but would then appear at the counselor's door to say they didn't belong there. The academic counselor quit the first week of September shortly after admitting that he "had no idea about the challenges of urban education."

It was clear we needed to achieve order and set behavioral expectations for the school day. Student behaviors were interfering with classroom instruction, and teachers were vocal about it. Teachers cannot teach if there is no order. Teachers veteran to the "original" school and its self-selecting, truly interested-in-science-and-mathematics student body were struggling under the newly declared open-enrollment policy that proclaimed, "We will educate your child and be a catalyst for their future—regardless of interest, demonstrated ability, or motivation."

Some of the teachers who were hired immediately after the storm and prior to my arrival were overtly empathetic and decidedly justice-oriented toward disadvantaged students. There was no "hidden" agenda for them; it was obvious. While empathy and teaching with a moral purpose are essential underpinnings of work with inner-city students—with *all* students—so too are firmness and fairness. Given the circumstances, too much leniency for repeat infractions meant that teenagers were not being held accountable. Rather than acknowledge and report the volatility of some students, the more empathetic teachers made exceptions for unacceptable behaviors and allowed students to leave the classroom and walk the halls unaccompanied. That permissive attitude flew in the face of the few, more experienced teachers in the building who deftly executed classroom management and set expectations for students to be in class, use impulse control, and respond to requests for classroom order and instruction with respect and acquiescence.

For certain students, disruptive behavior appeared to be a normal operating mode. Such behaviors often accompany a student's academic deficits, with frustrations manifesting as angry outbursts. It was evident that many students needed help beyond the sympathy of kind, educated, well-meaning teachers and administrators.

I found myself trying to be everywhere at once, walking the hallways and stairwells to redirect errant students back to class or to the social worker, available to intervene in classroom disputes between students or between teachers and students. Many of our teachers were barely 20-somethings, with big hearts, no teaching experience, and no experience working with educationally underserved inner-city youth. They were apt to confront students directly about offensive behaviors, rather than de-escalate a situation by lowering their voice or giving the student a chance to save face. Some students exploited the inexperience of these young teachers by taunting or simply ignoring them.

When one such student, who had been asked to leave the school after numerous suspensions, was shot and killed several months later, it threw everyone into grief, guilt, and pain. Another student, a freshman girl who attended school sporadically but always managed to make her presence negatively felt, was murdered in a drive-by shooting in the city's violent Central

City area. The legacies of poverty—violence, turmoil, and confusion—spilled over into the school on a daily basis. All of this was exceedingly draining on everyone. I knew that the kids needed help, the teachers needed help, and I needed help. Yet at the end of the day, none of us could make the neediness go away. The year ended with my hiring an assistant principal and deciding on a plan to implement the state-approved program called Positive Behavior Support (PBS) at the start of the next school year (see http://www.nasponline.org/resources/factsheets/pbs_fs.aspx).

The entire faculty worked together to establish measureable and achievable goals and brainstormed ways to make next year better. We wanted teachers to manage their classrooms; we wanted students to conduct themselves appropriately; and we wanted parent support. Ultimately, our goal was to create a comprehensive learning environment where students could grow academically, engage in activities outside of academics that sparked their interests, and have access to physical and mental health care.

## MAKING PROGRESS UNTIL...

At the beginning of the second academic year after the storm, the leadership team worked to implement PBS as a vital component of a comprehensive student support system. The dean of students worked directly with the new assistant principal to deal with behavior infractions and to work with parents to identify "trigger points" that often ended in students being ejected from classrooms. The new behavior program involved a tiered approach to discipline, starting with verbal and written warnings and ending with referral to the dean of students.

To reinforce the PBS program, a new teacher-mentor program was put in place that emphasized classroom management strategies, though admittedly, the few times we met were not enough. The overall goal was to keep students in class, in a supportive learning environment. Recognizing that at times students would need to be removed from a classroom, we provided a safe and controlled environment where students who needed a cooling-off period could go to collect their thoughts, write about them, and brainstorm better coping strategies. A room was set aside for this purpose and staffed by a designated employee. If it was necessary to suspend a student, we preferred to use an in-school approach, depending on the infraction, and teachers supplied students with daily assignments to help keep them from falling further behind.

We tracked student behaviors through PBS in combination with the Response to Intervention (RTI) program, which helps identify and respond to behaviors indicative of special learning needs (see http://www.rti4success.

org/). This allowed us to intervene before a student entered the self-defeating spiral of failing academically due to ejection from class and missed instruction and then becoming increasingly frustrated and destructive, only to be ejected from school. A significant number of students needed counseling support, and I hired a part-time, and then a full-time, social worker for the school through a grant from the Louisiana Institute of Mental Hygiene. A powerful and beneficial alliance was formed with the Louisiana State University Hospital. An adolescent health care clinic, initially housed in two rooms in the lower part of the building, received state and federal funding. Another grant enabled the transformation of an old and decrepit cottage on campus into a state-of-the-art, adolescent health care clinic. This facility provided important services for students, many of whom had not seen a doctor in years and did not have up-to-date vaccinations. Student medical needs could be addressed on campus, and those beyond the scope of the clinic were referred elsewhere. An additional social worker was hired, making a second full-time counselor available to our students. We shared the services of a psychiatrist who scheduled weekly visits with identified students. The clinic served many, many kids who, had they not come to our school, would have received no help at all.

The 2008–2009 year started off with a well-attended orientation for parents and students to explain the PBS system and make behavioral expectations clear. We wanted and needed parent support for our efforts and were working toward our goal to establish parent-school collaboration in promoting a respectful learning environment. As the opening weeks progressed, teachers remarked that students were polite, were generally responsive to requests for their attention during class, and the ethos of the building felt positive and energized. A new location for student services funneled students away from the main office and the counselor's room, providing a central location for managing tardy slips, clinic visit referrals, uniform checks, and similar functions. Increased cafeteria supervision, along with reorganizing the flow of traffic and food distribution in the cafeteria, helped process students more efficiently. The initial momentum did not last.

In late August 2008, a new and powerful storm, Hurricane Gustav, was forming in the Gulf. The tension across the city and in our small school was palpable. The counselor's office gave students a copy of their transcripts, stamped with the school seal, "just in case." The local school district that had authorized our charter distributed emergency cell phones with a central out-of-state contact number, a lesson learned from Katrina when all the cell towers were down and communication was silenced.

A steady drizzle ensued. The pulse of the school quickened and speculation abounded about what would happen "this time." Teachers were assigned an "anchor" class and asked to record students' contact phone numbers since family phone numbers on the emergency data cards were often incorrect or out of service. The kids had their own cell phones and ready access to each other. Teachers shared their own cell phone numbers with their anchor class so they could maintain contact, locate students, and then report in to a central number. The school phone rang incessantly with parents asking if we were going to dismiss early. Many parents of families living in low-lying areas susceptible to flooding had already come to retrieve their children.

As an independent charter school, we could make our own decision about closing. We were, however, attempting to coordinate with the district authorizer and maintain some consistency among schools linked to the newly formed East Bank (EB) Collaborative, an alliance of public charter schools in Orleans Parish. At around 1:15 p.m., I spoke with our governing board president and made the decision to close in conjunction with other schools in the EB Collaborative. Later in the afternoon, an evacuation order was discussed for the city for Sunday, August 31, as Hurricane Gustav strengthened and the city once again appeared to be the target. Many of the kids relied on public transportation, and a time had been set for service to be halted, so students raced down to the streetcar, some screaming excitedly and others hugging and crying.

Hurricane Gustav struck on September 2, 2008, shutting down the city for four days. Afterward, residents trickled back, and while we were spared the devastation of another Hurricane Katrina, the order and forward movement of the school year had been disrupted. It felt that we were starting all over again. Gustav revealed the level of underlying fear and uncertainty that remained in the psyche of the students—and the adults. I gained an appreciation for how quickly the rhythms of the school day, and life in general, can be disrupted.

### HELPING THE KIDS

Despite the challenges of starting a charter school under the most trying of circumstances, many successful programs and practices were put in place to benefit students. We created wrap-around services in a small, caring school community, a "one-stop-shop" where students could come to school to learn and receive academic and emotional counseling, as well as medical attention. In the afternoons, an extended-day program, staffed primarily by volunteers in conjunction with a program called Outreach, provided tutoring and extracurricular activities. Kids engaged in poetry slams, played basketball, and took classes in dance, drama, and keyboarding. A small group studied Chinese; another investigated and expanded their abilities in Japanese *anime* and *manga*. Teachers led special tutoring sessions for students who needed extra academic help. Students were assigned mentors who would check in with them throughout the week and encourage them to exceed their own expectations. We began a successful freshman-academy approach to build community and mutual support by assigning all freshmen to a team of teachers who could get to know them and target their needs.

Unique to our school was the requirement that freshmen take yoga for physical education, which provided them an opportunity to learn the basics of centering and meditation as well as physical stretches and positions. While few schools have tried this approach, we found it helpful, as students learned to control their breathing, focus on positive thoughts about themselves and the world around them, relax, and take those practices with them throughout the day. The course served them well, especially when they encountered a situation that required them to stop, step back, and think, before acting. The yoga teacher very successfully worked with students who never imagined themselves "takin' that yoga stuff."

To help coordinate student services, we created a school advisory team that met on a weekly basis to talk about what students were doing in class and to brainstorm ways to help those who were struggling. Academic and behavioral plans were written to address student needs. Follow-through was difficult because everyone had so much on their plate, but in spite of the challenges, we were able to reach many students and prevent them from becoming derailed academically and behaviorally.

We continually tried to elicit the help of parents, but they were struggling with their own problems of homelessness, joblessness, sleeplessness, and feelings of helplessness. The moms often came into my office and began to talk, but would dissolve in tears, sobbing, "I don't know what to do. I'm praying for my son to go to the counselor. I don't know what to do for my child—he's just not the same child, and I'm not the same either." While kids were in school, the adults were fighting with insurance companies or trying to secure work and get their lives in order. People readily admitted taking antidepressants with no talk-therapy or follow-up to help them learn coping skills. We weren't alone. Post-Katrina, mental health care was in short supply in the city, and any providers who managed to restart their practice were stretched beyond their capabilities.

## HELPING THE TEACHERS

In retrospect, I clearly see that more needed to be done to equip new teachers with strategies for working with students affected by the tumult that followed the storm. The youth and inexperience of so many of the teaching staff played out on a daily basis in calls for assistance. One of the teachers, a knowledgeable doctoral candidate at a nearby university, used PowerPoint and integrated video clips from a scientific series in his chemistry class. He had great success with those students but was defeated in efforts to teach students in his biology class. The less capable and less compliant kids detected and exploited the teacher's vulnerabilities early on. They challenged him and talked back. Lacking skills of patience and maturity, the teacher responded with threats of punishment he never carried out. Every day there was a crisis in his class, with someone being sent out of the room or someone having to go in to intervene and reestablish order. The kids began to completely ignore him. He left at Christmas that year.

A teacher who came to us through the Teach for America program also struggled. With the exception of a few focused students, the kids blatantly ignored his lessons, choosing instead to talk or laugh at disruptions. The dean or other staff could intervene and quiet the classroom for a while, but as soon as they left, the situation quickly deteriorated. The teacher began to stay at school very late, determined to prepare lessons to engage the students who would not give him the time of day. He began to direct his instruction to the two or three students in the class who were trying to learn, to pay attention. Together with the dean of students and the department chair, classroom interventions were routine. We took turns sitting in the class on a regular basis and meeting with him to help him strategize management skills.

The problems in this young teacher's class were not an exception. Out of 30 teachers on staff during the 2008–2009 school year, only 7 had three or more years of experience and understood how to establish boundaries, interact with students, and follow through with discipline in a firm but fair and consistent manner. Besides youth and a spirit of justice, the commonality among the new teachers was that none had been through a formal school of education and thus lacked the sustained learning opportunity to work with, observe, and be observed consistently by master teachers. This problem remains today.

The leadership team worked with individual teachers to help them get across to kids that they could control their behaviors, that they had choice in the matter of learning and behaving, and that they were making choices each and every moment. We worked with teachers to help them de-escalate potentially volatile situations. For example, we advised them not to block a door if a student was intent on leaving the class. Standing in a doorway invited a challenge, and the student would likely push past the teacher, resulting in greater disciplinary consequences for the student and increased agitation among others in the room. Common strategies for dealing with conflict included lowering the voice, stating that the student had a choice, telling the students they were welcome in class but must choose to obey rules, and calmly acknowledging their dissatisfaction and suggesting options: "Okay, I hear what you are saying. There are other people in the class trying to learn, so you can choose to stay and work with us, or you can go to the office and be suspended. You're making choices."

In one-to-one coaching sessions with various teachers, I would ask them to practice taking a step back under tense circumstances. By walking away from an agitated student, breathing, and letting a few seconds pass before responding to a student, the teacher can maintain both personal and classroom control and give the student a chance to "pause and re-set." In this way, teachers avoid taking the "bait." They demonstrate implicitly that they are the professional and the classroom leader without explicitly calling out a student in the moments following an outburst, confrontation, or disruption. In a challenging situation, I advised teachers to continue instruction, cross to the other side of the room, keep the rest of the class on task, and calmly buzz the office for someone to come and escort a student to the office. They were specifically advised against saying the student's name when calling for assistance. This omission allowed the student to save face and did not reinforce the student's acting out.

Most teachers—most people—think it reflects poorly on them when they have to ask for help, but teachers new to the classroom need to know that back-up is available and that the students often do not mean what they say personally. This isn't always the case, but my understanding of students who respond in a hostile or other inappropriate manner is that there is a larger issue than the one on the surface.

Many underserved students in inner-city schools have lived lives that young, suburb-raised teachers have only read about. I write this not to diminish the positive affect and empathy exhibited by so many teachers, but to emphasize that training, practice, and role play are needed to equip teachers with strategies and skills to respond in rational, helpful ways to children and young adults who have been victims of traumatic life-episodes. Such students often lack solid ground on which to stand, and they may lack calm, poised role models to imitate when faced with distress and frustration. They have witnessed rebellion and obstinacy in the face of upset or other challenges and are too young to respond reasonably.

These are common themes and responses in the lives of those who have been victimized. Lacking control over the basic needs of their lives—safety, food, shelter—they take control where they can and demand to be heard, regardless of how harsh and hurtful the effect on others.

It is critical, therefore, that teachers receive ongoing training, counseling, and encouragement from school leaders and organizations allied with the schools to promote the safety and well-being of students and educators alike. Leadership and teacher training programs need to provide information about how people respond in the aftermath of tragedies as well as appropriate, effective techniques to deflect, de-escalate, and validate those who have lived through such experiences.

## CHARTER SCHOOLS IN THE AFTERMATH

There was no map to help navigate the educational landscape in New Orleans post-Katrina. With the district virtually dissolved and a state mandate for change, restarting education was an uncharted and entrepreneurial endeavor. Even formerly established schools that converted to charter status after the storm were feeling their way through, searching for teachers, for food service providers, for custodial services, for technology sources and support, even for ways to attract, register, and enroll students in classes. With few exceptions, each charter school was unique to the vision of its governing board, and structurally and systemically experimental in nature. As a result, we were literally "building the plane while flying it."

Inherent in the charter school concept is the emphasis on accountability and individuality. However, there is no one management prototype. Some charters are organizationally structured as traditional pyramids, while others rely on co-directors who divide duties. Some have boards that hold tight reins and require fastidious reporting from the director; others give directors carte blanche decision-making and managerial trust. Some charters budget for operations and have finance managers; others leave such work to the director.

Although the idea of autonomy appears liberating, charter school educators face additional demands because the director and staff must also serve as the central office. As a result, while this design allows for experimentation and on-the-spot decision making that can directly benefit students, it also carries with it an amorphousness and ambiguity to which not all are suited or equipped from a professional knowledge or training aspect.

Anyone considering taking an administrative position in a start-up charter school that has experienced a natural or man-made disaster should ask question after question about the scope and depth of the expected leadership duties. The multiplicity of tasks and the infinite number of needs I tried to meet give me pause. It was an around-the-clock job in which I worked with the governing board to strategize the *what, how,* and *where* of compliance issues, building/facility upkeep, funding sources, and the concept and possible location of a second school, while managing immediate school complexities of staffing, curriculum, discipline, instructional accountability, charter-authorizer requirements, and student and parent needs and concerns. To give you an idea of what this involved, consider the following sample of duties I, like other charter school principals, routinely performed as we brought schools back to our devastated city:

- Worked diligently to attract competent teachers to build a staff reflecting experience, energy, and know-how.

- Completed a School Improvement Plan to improve students' scores and identify intricate uses of Title Funds.
- Met with teachers and special education providers to organize the delivery and oversight of services for students with exceptionalities.
- Together with the dean of students, wrote, presented, and received governing board authorization for the student and faculty handbook.
- Met with parents, students, and teachers on disciplinary matters and participated in numerous expulsion hearings.
- Established a salary scale with bonus options for hard-to-hire placements.
- Worked with financial officers to review, reshape, and devise creative ways to stretch the limited budget.
- Attended countless meetings called by the charter authorizer with minimal notification, always requiring the rescheduling of other work.
- Met with the foundation director and board to strategize ways to raise funds to support the school's mission.
- Sought ways to get the leaking roof repaired while praying that plaster would not fall on the heads of students, thereby causing injury and generating a law suit.
- Traversed the city looking for new school sites for a planned second school, as well as for a more suitable site for the school I was leading.
- Dealt with both trivial and significant schoolhouse politics, including attending a city zoning board meeting in an attempt to ward off a change to the school's footprint.
- Implemented and followed through on the assistance and assessment program for training and evaluating new teachers—an exercise characterized more by burdensome paperwork than by authentic teacher training and assistance.
- Interviewed, counseled, hired, and fired personnel for reasons of incompetence and criminal activity revealed after fingerprinting results finally became available (five months into the school year).

Duties of school leadership extend beyond the physical school grounds, and I found myself fielding calls from neighbors about student behaviors and meeting with neighborhood association representatives to foster goodwill and address concerns. Together with a few teachers, I directed traffic in front of the school because parents engaged in screaming obscenities at each other. Drivers ignored street rules and parked anywhere they wanted to pick their kids up, even if it meant going against traffic and endangering the lives of students. On many occasions, as the school day neared its close, I would walk down to the streetcar stop to talk with transportation officials who refused to let our students on board because some had previously behaved inappropriately. While there indeed were obstreperous and out of control kids, there were also many students who understood what it meant to wear the school uniform with pride and to represent themselves and the school with respect.

School leaders, especially after an occasion of disaster, need access to professional development and counsel. After Katrina, one organization began fostering charter school leadership training in the city. Their "grow a leader" approach involved providing selected candidates a year of training, with site visits to successful charter schools around the country. School leaders would then be assigned a school and "grow it" by adding one grade at a time. While this is good way to cultivate a school leader who could then help shape a school culture, the reality is that there were more charter schools in New Orleans than there were available school leaders. Colleges and universities could address this problem by expanding their focus in licensure programs to include leadership not only in traditional school settings but also in public charter schools. In order to understand the responsibilities, would-be school leaders could benefit from three- to six-month apprenticeships in different school settings—charter and traditional public, as well as urban and suburban.

Efforts in New Orleans have been under way for several years now to create more centralized support for charters through collaboratives or associations. Additionally, charter organizations, such as the Knowledge is Power Program (KIPP) and new entrants in the charter movement have expanded the number of schools they oversee based on models they developed. They have created mini-centralized support offices that assist in providing and overseeing teacher training, curriculum writing and monitoring, technology set-up, data-driven decision making, as well as software purchasing, food preparation and provision, and custodial services. While smaller schools (350–400 students) offer a more personalized educational setting and services, it is difficult to achieve an economy of scale, and redundancies abound in administrative, purchasing, and supply costs. Collaborative alliances have the potential for resolving some of these obstacles.

### FUTURE-FORWARD THINKING

Seven years after the storm, the city of New Orleans and surrounding environs affected by the storm of a lifetime still bear the watermarks and other evidence of destruction wrought by the levee breaches. Still visible are the enumerated scrawls left by National Guard units as they swept through houses searching for survivors in the months after the storm. Waterlogged, dilapidated, and unusable school buildings linger as reminders of neglect and educational failure. The populations of newly chartered public schools continue to be housed in mobile units that often reek of unsanitary odors, or are crowded into the few remaining buildings alongside existing school programs.

There is, however, growing confidence in the ability of charter-operated schools to make a difference in the learning and in the lives of students as demonstrated in increased student academic performance on standardized tests across grade levels in charter schools throughout the city. An anomaly in public education in the United States, charter-operated schools now dominate the education landscape in New Orleans, offering choice in a city where robust, beneficial change and action are new players.

In 2010, the New Orleans public school student population was 40 percent less than what it had been before the storm. Over 70 percent of these students were attending a charter school of choice. New Orleans had become the most charter-operated, school-choice city in the nation. I am not suggesting that charter schools are the answer to educational transformation, but in New Orleans they have become a way to transform lives. Achieving the type of educational change that matters means providing students the education they must have in order to become learners, thinkers, problem solvers, caring individuals, and future leaders.

In the aftermath of Katrina, I have seen students, educators, and community members show enormous resilience and courage in the face of uncertainty, hardship, deprivation, and demoralizing conditions, with the bare minimum of supplies, in moldy, visibly unhealthy buildings with rotting floors and windows barely secured in their casements. Having worked in public schools in other locations, I know the kinds of buildings, resources, and working conditions that other students and teachers have available to them. Hurricane Katrina exposed harsh inequities to the world, and many people have taken action to help in the clean-up and rebuilding effort. However, the legacy of years of neglect will take decades to remedy and improve.

There will always be disasters, but there will also always be small victories and shared successes. I want to close this chapter with a story of one of those successes.

Several weeks after Hurricane Gustav forced the evacuation of New Orleans in 2008, I received a letter from a woman in the Midwest. As I opened it, a check fluttered out along with a note. The note explained that the woman and her husband had been vacationing in New Orleans when they met some of our students being sent home early due to Gustav's evacuation mandate. She said they had been riding on the St. Charles street car when

a large group of students wearing your school uniform boarded, and we were struck by their conversations. They were visibly, vocally upset by the school's closing and the impending evacuation. They animatedly expressed and shared their fears with us when we inquired what was upsetting them. "Not again"—"Where will we go?"—"We can't miss school"—"We can't lose our friends again"—"We're just getting back on track."

The woman went on to say, "We were so impressed by their articulation and so moved by their emotions and sincerity that upon returning home, we held a golf tournament and raised some money that we hope you can put to good use on their behalf."

This is just one example of strangers taking notice of the promise and the needs and wants of our students. Compassionate and generous gestures continue to surprise us. Newcomers and locals, eager new teachers and veteran educators alike, have responded to the need for transformational education here in our city following the tragedy. We must dedicate ourselves to unceasing work and love so that the Katrina-kids can eventually come to flourish in the aftermath.

*   *   *

**CONTRIBUTOR NOTES**

**Bridget Dwyer Ramsey, Ph.D.,** former director of the New Orleans Charter Science and Mathematics High School and the Academy Charter School, Colorado, has over 18 years of experience as a teacher and educational leader in public and charter high schools in both suburban and urban settings. She has presented, researched, and written about high school reform and charter school development and sustainability.

**WHAT'S NEXT?**

The New Orleans public school you will read about in Chapter 6 had developed plans for becoming a charter school even before Hurricane Katrina hit and, as a result, had a head-start on resuming functions. Through a partnership with Tulane and other private universities in the city, Lusher School was resurrected to provide an education for the children of returning university faculty and staff. Science teacher James Whelan describes the logistical and curricular strategies that facilitated the reopening of this school.

# CHAPTER 6

# "FAKING IT 'TIL MAKING IT"—A KATRINA STORY

## JAMES A. WHELAN

BEFORE HURRICANE KATRINA, LUSHER SCHOOL WAS A HIGH PERFORMING magnet school for students in kindergarten through grade 8 (K-8). After Hurricane Katrina, it became a K-12 charter school, in partnership with Tulane University. This chapter explores how a disastrous storm drove this change, how the storm affected faculty and students, and how they adapted to their changing world.

I teach science at Lusher Charter School, and much of this chapter comes from my personal perspective. To provide other viewpoints, I conducted a series of interviews of fellow teachers and administrators. I asked them to share with me—so I could share with you—their experiences, their challenges, the changes that Katrina brought to schooling in New Orleans, their thoughts about barriers to reopening school after the storm, and their recommendations for moving from triage to teaching after the storm.

### BACKGROUND: LIFE BEFORE

Lusher opened as an elementary school in 1916 in uptown New Orleans on Willow Street, not far from the campus of Tulane University. In 1990, a separate middle school was formed nearby in the old court house on Carrollton Avenue. Both the middle and elementary schools have had rigorous academic programs complemented by a rich arts-based curriculum. Admission to the schools has been both by application (for most students) and open to all in Lusher's attendance zone. Parental involvement has been high, community support strong, and the school's research-based approach to literacy and numeracy instruction has produced achievement scores that are among the highest in Louisiana.

Lusher has a well-established set of traditions, ceremonies, and celebrations, and the Lusher community has maintained continuity through a set of guiding principles called the "Project Pride Rules." The rules are quite simple: *Be kind; Be respectful of people and property; Be responsible;* and *Do your best.* The student population of Lusher has always been diverse, closely approximating the diversity of the city. Yet, through Project Pride and the arts program, the school has built a homogeneous culture, The *Lusher Way.*

In the months prior to Katrina, the Lusher school community had decided to form a charter school, with parents and teachers alike voicing overwhelming support of the plan. On Saturday, August 27, 2006, much to the displeasure of their loved ones worried about evacuating—the school's leadership team worked until 11:00 p.m. to complete the charter applications and submit them electronically both to the Orleans Parish School Board and to the Louisiana Department of Education in Baton Rouge. The School Board was expected to consider this application at its next meeting, scheduled for the first week of September.

When classes were dismissed on Friday, August 26, Hurricane Katrina was a large storm entering the Gulf of Mexico with a projected path to strike Florida's Gulf Coast. By Saturday morning, it was clear that coastal Louisiana would receive at least some damaging winds, rain, and high tides from the storm. South Louisianans went into their usual hurricane response routine. For people in the lowest areas, that meant packing up and getting out early, before rising waters blocked the roads. For people in the city, it was pick up and batten down everything outside, make calls to relatives or friends seeking a room for a night or two, or if necessary make a hotel reservation somewhere out of harm's way. Evacuation typically involves driving out of town in a big traffic jam, spending a night or two jammed up in some relative's house or hotel room, followed by another traffic jam return once the all-clear has been sounded. As the day progressed and the forecasts became more ominous, the city people began to leave.

By Sunday, Hurricane Katrina was powering up as it headed across the Gulf, with winds at one point estimated at 175 mph. The City of New Orleans ordered mandatory evacuations and designated the Super Dome as a shelter of last resort for those unable to leave.

Everyone expected to receive an all-clear after a day or so and then return home to tend to a few broken tree limbs or lost shingles. Instead, we watched unimaginable devastation for weeks on every newscast. For many people, weeks and months would pass before they could return to even assess the damage to their homes and places of work.

Within a week or two, residents with a home that had not been flooded were allowed back to survey the situation. Many of the fortunate ones were able to occupy their home, at least in some primitive way, within six to eight weeks. This usually involved a form of camping out. Potable water, electricity, and gas service were slowly returned across the city. Food and gasoline were in short supply. Many people split their lives between a community with services and New Orleans. Going to another location for food, fuel, ice, water, and building materials, they would return to their home for a day or two, make what repairs they could, and when supplies ran out, head back to restock and refresh.

The full extent of damage from Katrina may never be known. Federal reports estimate that during the storm, more than 1,500 lives were lost; approximately 770,000 people were displaced; and in the aftermath, nearly 275,000 Gulf Coast residents were forced into shelters. Some who safely evacuated never returned. Almost 80 percent of the City of New Orleans was flooded. Victims lost their homes, places of employment, churches, communities, and basic necessities. They also lost much-needed documents such as birth certificates, property deeds, proof of insurance, medical records, and almost certainly school records (e.g., FEMA, 2006; White House Archives, 2005).

### RESURRECTING EDUCATION

During the earliest stages of the recovery, the Orleans Parish School Board announced that they did not expect to open any schools until the following school year. Then they fired all teachers and school administrators. Because of the abysmal performance of many Orleans schools even before the storm, they had been under threat of state takeover for some time. In November, the state took action, establishing a Recovery School District (RSD) to assume control of the lowest performing schools in Orleans Parish. As a result, over 100 schools were transferred to the RSD for an initial period of five years.

At the time, Lusher faculty were scattered across the country. With the firing, they lost health insurance, accrued leave, and their jobs, thereby becoming eligible for unemployment compensation and food stamps. Most started looking for a job in the area to which they had relocated. Some returned within three or four weeks of the storm, but for many it would be January or even the following summer until they could return to the city to face many months of rebuilding before they were finally *home*.

These experiences were not unique to Lusher's faculty. Everyone in the city experienced some level of job insecurity, an unending set of challenges to rebuild, find temporary housing, and deal with the depression caused by all of the reminders of what was lost.

In spite of the great loss, a spirit and drive to rebuild began to grow among New Orleanians and across the country. Lusher's principal was one of the early leaders trying to get the schools running. Immediately after the announcement that schools might not open for a year, she took the charter application that had been completed before the storm to Baton Rouge and met with state education leaders, trying to find a way to restart Lusher. They offered every assistance at their disposal, but additional funds were needed.

As often happens, the outcome was shaped by someone with a different, but compatible, need, in this case, the president of Tulane University. Tulane, the largest private employer in the city, had significant flood damage both to its main campus in uptown New Orleans and to its medical school and hospital located downtown. The president wanted to reopen the university in January and needed a school for the children of the faculty and staff to attend. A partnership was worked out: Tulane provided the additional funding Lusher needed, and Lusher provided spaces for the children of Tulane's returning faculty and staff. This agreement was expanded to include the children of faculty and staff of all four private universities in the city. With the approval of the application to operate as a charter school, plans for a January reopening began to take shape.

By late October, the Lusher leadership team had re-formed. In less than three months, they needed to revive two disaster-damaged facilities, locate and recruit a faculty and student body spread across the country, and procure everything necessary to open and operate a school. Much of what they were attempting had never been done before. As a charter, they no longer had a school system to provide accounting, janitorial, and food services. All of this had to be accomplished in an environment where just buying a postage stamp was a challenge. Compounding the problems was the critical task of determining how to meet the emotional needs of students, faculty, and staff.

Leadership team members who participated in this massive task of preparing the school for reopening described it as a magical time. The hours were long, the conditions were terrible, and many of the tasks appeared to be impossible, but all were connected and everyone was willing to lend a hand regardless of the task. When someone had that "empty" stare, a colleague would bring them out of it. A survivor spirit and sense of community fueled this group to complete the impossible task of opening Lusher School grades K-10 by mid-January 2006.

Plans for Lusher can best be described as a shot in the dark. How many former students would be coming back? How many new students would be added? How many faculty would we need and how many would return?

Each question had a myriad of possible answers. Even when a decision was made, conditions could change. One faculty member trying to return from Shreveport pulled off of the highway in Baton Rouge and turned around. She could not deal with all of the unknowns and uncertainties of moving her family at that time. She was able to return six months later.

On top of all the questions to resolve before reopening the elementary and middle schools, a whole new high school had to be created. A core faculty of five was hired for the high school, one teacher for each academic area: mathematics, English, science, social studies, and foreign language. This core was supplemented by the arts teachers who would be teaching in the K-8 classes. The middle and elementary schools' faculty was composed of returning Lusher teachers. Because of the limited number of returnees and lower number of students, many of these veterans were shuffled around to new assignments. A week before the opening, most of the faculty had been hired, and final preparations were under way. However, some positions could not be filled until the weekend before the opening.

The first floor of the elementary school building had been flooded, and some classrooms received rainwater damage, but for most teachers it was close to what it was like before. Elementary teachers returned to the classrooms, materials, and supplies that they left behind in August. The middle school building now housed both the middle school and the new high school, with supplies and equipment being shared. Rooms were reassigned to accommodate the five new teachers and new students. The veteran staff of the middle school had to move to less desirable classrooms, share materials that they had collected and protected over time, and deal with a whole new group of students who were older and much larger than their own students. As a newly hired high school teacher, I can say that to a person, the middle school faculty was caring, generous, and patient toward all new faculty. From them, we quickly learned the Lusher Way.

On January 17, 2006, Lusher reopened its doors to 484 elementary students (grades K–5), 312 middle school students (grades 6–8), and 47 high school students (grades 9–10), for a total enrollment of 843 students. During the previous year, 1,157 students had attended Lusher in grades K-8 (Table 6.1). One grade level would be added to the high school each year so that in two years Lusher would offer a full K-12 curriculum.

Students who returned in January had experienced challenging situations over the previous four and a half months. Some had survived the horrors of the flood and observed the destruction of almost everything in their lives. Others had evacuated safely and were able to restart their lives within a week of the storm. Some returned to a home with little or no damage. Others would not return to their pre-Katrina homes for years, if at all. Temporary housing options included downtown hotels, FEMA trailers, and camping

Table 6.1    Lusher Student Enrollment

| School Year | Elementary School | Middle School | High School | Total Enrollment |
|---|---|---|---|---|
| 2004–05 | 652 | 505 | n.a. | 1,157 |
| (pre-Katrina) | | | | |
| Jan–June 2006 | 484 | 312 | 47 | 843 |
| 2006–07 | 624 | 413 | 216 | 1,253 |
| 2007–08 | 649 | 428 | 355 | 1,432 |
| 2008–09 | 655 | 422 | 392 | 1,469 |
| 2009–10 | 654 | 449 | 417 | 1,520 |
| 2010–11 | 712 | 473 | 451 | 1,636 |

out in an undamaged portion of a flooded home. Families with a livable home filled every spare room and sofa with relatives or friends who were less fortunate. Some students commuted from outlying areas over an hour away. Some families chose to temporarily separate: One parent would stay with the younger children in the stable community to which they had evacuated, and the other would return to New Orleans with older children. In some cases, a child was sent to live with a friend or relative in order to come back to Lusher.

Making suitable living arrangements was only one of many issues that had to be addressed as people struggled to recover. Transportation was another. Almost every car left in the city during the storm had been ruined by the floodwaters. Finding a replacement vehicle was a major challenge, and public transportation was intermittent at best and nonexistent for much of the city. Added to this was the fact that the city's infrastructure had been compromised and was only gradually being restored. Each day provided a new set of challenges: finding a doctor, dentist, or barber. What pharmacy was open and when? The pre-Katrina support system of grandparents, relatives, or friends was broken. Many neighborhoods averaged only one or two occupied houses per block.

Living in this unpredictable, isolated community produced frustration and stress at every turn. Depression was the normal state of mind for many people during this time. With mental health services almost nonexistent, many attempted to self-medicate with alcohol or whatever drugs were available. For several families, divorce was added to this toxic list of stressors. Amazingly, most of our students arrived with a positive attitude ready to start back to school.

In the aftermath of Katrina, lack of contact information and loss of records and documents were major impediments to restarting school. To be better prepared in the future, contact information was collected for all students and faculty and stored along with important school records on hard drives located far from coastal Louisiana. An Internet home-page was designed to provide a central communication site that would simplify reconnecting with families and school personnel. (In August 2008, this crisis plan helped ease the evacuation and recovery from Hurricane Gustav, when students were given copies of their transcript prior to evacuation, and the contents of every room were videotaped so losses could be documented.)

## TEACHING AFTER THE STORM

Psychology professors from Tulane and Walden University, the school social worker, and the leadership team created a *healing curriculum* designed to help students develop coping skills and build peer support through small group activities and teacher-facilitated discussions. A block of time was set aside each week for students to participate in this experience. University professors provided training and follow-up discussions for teachers, and several doctoral students from the psychology department provided additional support. Students with greater needs were identified and referred to the school social worker. The healing curriculum was used from January through May of 2006. Elementary and middle school students responded well to the program, but high school students were less inclined to participate.

Developing a plan of instruction for students living in such difficult conditions was another challenge. While New Orleans schools were closed, some students had attended high achieving schools, others were in poor rural or urban schools, and some were not in school at all. Many students had attended several schools over this period. Every class period presented a mixture of students, some having mastered most or all of the concepts, others lacking any familiarity with the concept whatsoever, and the rest somewhere in between.

The elementary school operated on a regular schedule with teachers working within their class or grade to fill in gaps and bring students up to grade level. Middle and high school classes were arranged in four double-periods, or 100-minute blocks. Each student was assigned three academic classes and an elective for each quarter. Middle school teachers used strategies similar to the elementary teachers to bring all students up to level. A special summer academic enrichment program was offered for all students who were below grade level by the end of the second quarter.

High school posed a different dilemma, for students needed to complete Carnegie Units, and thus lacked the flexibility of the lower grades. (A Carnegie Unit is a standardized measurement used to document completion of courses required for college entrance.) To ensure that students earned sufficient credits, during each quarter of double classes, students would complete one semester (a half-Carnegie Unit) of work. Students who had completed fall semester at another school would complete the second semester of *half* of their classes each quarter. Students not in school during the fall were able to complete four, full courses over this two-quarter period. They made up the missing courses during summer school.

On paper, all of these plans sounded fine. The reality was that each elementary and middle school lesson had to be designed to meet at least three levels: students new to the concept, students with limited understanding and in need of guided practice, and students at the mastery level needing to do higher level activities or move on to the next concept. In the same high school science class, for example, I taught both chemistry and biology with some students in each subject doing first semester material and others doing second. Fortunately, we had fewer students, but the widespread range of students overwhelmed that advantage.

In spite of these concerns, school had opened, students had a schedule, and they were in class. From day one, Lusher looked and acted like a normal school. There were many bumps in the road, but for seven-and-a-half hours each day, students were in a comfortable, caring environment. School was an island of normalcy, an escape from the post-Katrina chaos of New Orleans.

Of course, there were many limits to this normalcy. Assigning homework was difficult if not impossible. Drug and alcohol abuse emerged in the high school. There were ongoing difficulties with Internet access and limited help for computer problems, which was especially challenging since many students depended on the school for Internet access and computing needs. The school provided a daily meal, but damage to kitchen facilities and limited food supplies resulted in many less-than-appealing meals. Funds were limited during this half-year, so teachers had to be especially frugal and creative in order to acquire needed materials. The next school year, when federal money came pouring into the local schools, the situation changed significantly.

In the class-structured city of New Orleans, this was a time of little or no class distinction, a feeling reflected in the way our students interacted. At lunch, students mixed freely. There was a strong sense of caring for one another. A student struggling with class work always had a peer offering help. With few forms of afterschool entertainment available, students collaborated to create their own. Every student was involved in one of the art classes, which often provided a common ground for socializing.

Returning to school gave students a chance to feel something like their pre-Katrina lives, with a normal daily routine, the cycle of the school calendar, and the usual celebrations along the way. School provided a place for students living in isolated parts of the city to gather and socialize. By talking with their peers, they discovered that they shared many of the same hardships and joys. They reconnected with old friends and made new ones that may not have been accessible to them in pre-Katrina New Orleans.

## SCHOOL YEAR 2006–2007

The second academic year after the storm, a number of external forces impacted Lusher School. The middle and high schools moved into the building previously occupied by a low-performing Orleans Parish high school that was not reopened post-Katrina. Families continued to return to the city in increasing numbers. Vast sums of federal money were being spent to refurbish all of the schools, and rebuilding efforts were in full swing except for the hardest hit areas. This building boom brought new workers and their families into the city.

While significant progress was being made, new struggles were developing. In 2005, the Louisiana Legislature had empowered the Louisiana Department of Education to take over the low-performing schools from New Orleans. To accomplish this, the Louisiana Recovery School District (RSD) was formed, and they immediately took over all but 12 of the public schools in New Orleans. The Orleans Parish School District (OPSD) opened and operated 4 schools since 2006 and issued charters to 12 schools. In 2010, the RSD operated 25 schools and had issued 55 charters. The confusing and overlapping management of public schools that resulted from the two school districts and all of the different charters reached its zenith during the 2006–2007 school year. The OPSD's elected board was almost powerless and bankrupt. The RSD was trying to manage a local school system from Baton Rouge using the state bureaucracy. To these competing and confused management structures, large sums of federal dollars were coming in that had to be spent within a short timeframe. What followed were fights over the spending of every dollar. In addition to these two agencies were other entities in charge of clean up, waste removal, inspections, and permits. A school administrator trying to open and operate a school in the aftermath of Hurricane Katrina faced challenges on every front.

Lusher Elementary School continued to operate from the same campus, growing by 140 students from the previous spring. It was able to remain on the fringe of administrative chaos described above. From a student's perspective, it would have been difficult to distinguish that school year from most others.

The middle and high schools were not so fortunate. The four-story building they were moving into was built in the early 1930s and at its peak served 2,000 students. The physical plant had been neglected for decades. It had minor flood damage on the first floor and wind and rain damage throughout. It was a fully equipped school that had been shuttered after Katrina. As a precaution, after the storm all of the furniture and equipment had been moved to the interior hallways. Over the spring and summer, our leadership team inventoried these items and made plans for using the building.

Among the first problems encountered were the questions of when would the building be turned over to Lusher; who needed to inspect the facility; what permits were required; what equipment was usable; what was OPSD property; and what belonged to the RSD. The opening of school was planned for mid-August, with the week prior set aside for teacher professional development and setting up of classrooms. Far from the orderly process that the leadership team planned, chaos and crises ensued. During the first week of August, someone in authority declared that there was a potential mold problem and that all furniture, material, and equipment in the building had to be removed. For the next two weeks, the school's administrators wrestled with contractors, inspectors, and the fire marshal, and finally received approval to use the building the weekend before the school opened. During the week prior to the opening of school, faculty attended a workshop at a downtown hotel. On Friday afternoon, they were allowed to enter the building and begin setting up their classrooms. Over the weekend, parents, community volunteers, and teachers worked frantically to make the facility presentable for Monday morning's opening. The hastily painted walls were still tacky to the touch, and the custodial crew had worked through the night to clean up after volunteers departed.

On that Monday morning, the students entered a facility that looked like a school. It was far from the facility we wanted, but it would work and it has been improved over time. Every class had a basic set of furniture but not much else. There were two working restrooms for the entire school. But by 8:30, every student had a schedule and was in class. Of course, there were adjustments that had to be made, but we were a school.

Shortly after the opening, we began to regularly receive truckloads of furniture, equipment, and books. Again, someone in authority decided what we needed and shipped these truckloads to us. Most of the furniture and equipment were standard size and would fit into our classrooms; however, some large computer desks overwhelmed the space. Of the boxes of textbooks that were sent, some still remain unopened. Almost every classroom received three desktop computers and a printer. When these were hooked up and online, the school made a giant technological stride. Each teacher received a large box of office supplies, including a lifetime supply of sticky-notes. Huge amounts of money were being thrown at the problem, and teachers and school administrators had little choice but to receive the materials and use them in the most effective way possible.

The first few weeks in this building did provide a few challenges. For example, it was necessary to haul water up a flight of stairs to the chemistry lab, but within two months the lab had a sink and running water. There were no kitchen facilities, and lunches had to be brought in from another school. Lunch was often cold sandwiches. The gym was not usable but was renovated during the summer of 2007. Four different sports teams were fielded that year—swimming, cross country, volleyball, and boys' basketball. There were few other extracurricular activities. The routines of middle school were quickly established and adapted to the new facility. Their afterschool and sports programs were up and running within the first six weeks of school.

By the fourth week of school, things were operating smoothly. The new students had been folded into the school and were learning the Lusher Way. The procedures for operating in the new facility had become habitual. The first activities and traditions of the high school were being established. The first Junior Feast was held in November. Students in English III completed a unit on the literature of food and celebrated with a white linen feast that they had prepared. The Feast has been held every year since. With only grades 9–11 and the limited number of sports and extracurricular activities, Lusher was not a complete high school, but was well on the way to that level. Every week brought a new improvement, more restrooms and water fountains, broken windows repaired, new white boards replacing the faded chalkboards, and Ethernet connections for the computers that reached many classrooms.

In early January, tragedy struck. A popular 10th grade student who had been deeply troubled since Katrina died of a drug overdose. Several students were involved in the criminal investigation that followed. Her suppliers, two former Lusher students, were sentenced to the penitentiary. The entire school was shaken by this event.

The school responded to this tragedy on several fronts, including a thorough review of the school's drug policy and assessment of what was being done to educate students. As a result, a new peer-to-peer program for developing positive life choices was implemented. The school provided related training and resources for parents. Since that tragedy, whenever a drug-related problem has occurred, it has been quickly addressed with coordinated action among the school administrators, parents, and law enforcement. Community resources have been identified to provide counseling and treatment for students with substance-abuse problems.

In February, following the student's death, the school social worker and the psychology professor from Tulane who had developed the healing curriculum conducted a survey of stressors that middle and high school students were facing and how well they were coping with them (Overstreet & Petrosini, 2007). The survey was in two parts: The first surveyed the student's living situation and the degree to which various stressors were bothering the student; the second addressed symptoms of stress and trauma. The following instruments were used:

- For grades 6 and 7, Mood and Feelings Questionnaire-Child Version (Angold & Costello, 1987) was used to assess depression symptoms during the previous two weeks.
- For grades 8–11, the Los Angeles Symptom Checklist was used (Foy, Wood, King, King, & Resnick, 1997). This consisted of 43 items that provided a general index of stress, including 17 items that assessed post-traumatic stress disorder.

The survey of Lusher students was conducted 18 months after Katrina, a time when most of us thought things were getting back to normal. However, results showed that *normal* for students in New Orleans was still far from the normal experienced by most students elsewhere in the United States. For example, among the results were the following:

- More than 25 percent of the 8–11 grade students were bothered by at least five of the stressors measured.
- Approximately 25 percent of the 6–7 grade students were bothered by at least three of the stressors measured.
- Crime in the city impacted more than 50 percent of both groups.
- Parental stress was at nearly 40 percent for both groups.
- Over 30 percent of the students still lived in a damaged home.
- Approximately 35 percent had family members still away.
- Over 20 percent had others living with them.
- Nearly 40 percent had moved three or more times over the previous 18 months.
- Analysis of the second part of the survey indicated that 25 percent of the grade 6 and 7 students exceeded the clinical cutoff for depression, and 25 percent of the 8–11 grade students exceeded the clinical cutoff for post-traumatic stress disorder. (Overstreet, Salloum, & Badour, 2010)

As the school year drew to a close, improvements to the facility accelerated. Renovation of the gymnasium began. Planning for a summer renovation of the middle school science classrooms was underway. The loosening of control over funds for equipment allowed classroom teacher input into the decision-making process.

Other signs of improvement were becoming apparent. The student newspaper was recognized as the best in the city. For the first time Lusher had students taking Advanced Placement exams in English, Physics, and American History. A football team was formed for the next year, and the addition of soccer, track, softball, and baseball completed the athletic program. More clubs were starting up. Developments such as these took on added significance, since athletics and extracurricular activities gave students outlets for some of their stress while moving a few steps closer to a *normal* high school experience.

## 2007–2011 AND BEYOND

Year Three was a turning point in Lusher's post-Katrina progress. Up to this point, the approach of most of the faculty was to survive until next year. Pretend that things were normal, but wait for it to get better, or as one veteran put it, "faking it 'til making it." The impact of the storm was still present. The survey illustrated how much our students were still suffering; yet, over the course of the year, teachers reported a decrease in the Katrina stories that students shared. Instead of hearing one or more stories a day, they heard a Katrina story only once or twice a week.

A chemistry project assigned every fall offered anecdotal evidence of this change. Students were required to design a new 50-cent coin for this project. The design included selection of metals for the coin and images for each face. Students had to make a model of the coin and write a rationale for their selection of images and materials for the coin. In 2006 and 2007, themes of Katrina and rebuilding New Orleans predominated. In 2008, there were still more Katrina-related themes than any other but fewer overall. The themes ranged from a favorite baseball player to unicorns. By 2009, only a few Katrina-related themes, fewer than one per class, and President Obama was the dominant theme. In 2010, there were no Katrina-themed coins; instead the Super Bowl Champion New Orleans Saints was the dominant theme.

Katrina may no longer be foremost on the minds of Lusher's students but it is still there. Recent events illustrate this point. During a severe thunderstorm, a sixth-grade student became so terrified that she had to call home. In 2010, a graduating senior celebrated the return to her pre-Katrina home. Infrequently we hear a Katrina story at the school, but it only takes a little prodding to get students started.

The stressor survey uncovered how deep and widespread the effects of this disaster were on the students of Lusher School. Unfortunately, no follow-up survey has been conducted to determine its lingering effects. From anecdotal evidence, it appears that most students are doing fine with little or no residual damage, but it is hard to know. Houses have not been rebuilt, and some families broken apart during and after the storm have not been reunited. The collateral damage of drug and alcohol abuse following the storm has scarred many. The rebuilding of support networks to address these problems has been inconsistent.

As described earlier, the school has learned from the disaster and made many changes to better handle the next storm from an information management and communication perspective. Perhaps more importantly, we learned about the major role played by a school in rebuilding the social fabric of a community. Lusher School became a central feature in the lives of its students. It was one of their first experiences of returning to something that appeared to be normal. The routines, ceremonies, and traditions reassured students and faculty that they could get back to something called normal. Being with peers to talk and share experiences made the abnormal post-catastrophe experiences seem less strange. The school also provided parents with the opportunity to gather and share. Just waiting to pick up kids after school allowed parents a break from the isolation of their own recovery efforts and a chance to commiserate with others.

Seeing the school in operation gives a feeling of comfort to the whole community. It is a success, something is working, and it reinforces why we need to rebuild.

## REFERENCES

Angold, A., & Costello, E. J. (1987). *Mood and Feelings Questionnaire*. Developmental Epidemiology Program. Durham, NC: Duke University.

FEMA. (2006). Katrina one year later. *Federal Emergency Management Agency, U.S. Department of Homeland Security*. Retrieved June 27, 2011, from http://www.fema.gov/

Foy, D., Wood, J., King, D., King, L., & Resnick, H. (1997). Los Angeles Symptoms Checklist: Psychometric evidence with an adolescent sample. *Assessment, 4,* 377–384.

Overstreet, S., & Petrosini, A. (2007). Report of Lusher Middle and High school needs assessment. An unpublished report by Lusher Charter School.

Overstreet, S., Salloum, A., & Badour, C. (2010). A school-based assessment of secondary stressors and adolescent mental health 18 months post-Katrina. *Journal of School Psychology, 48,* 413–431.

White House Archives. (2005). Katrina in perspective. Retrieved June 27, 2011, from http://georgewbush-whitehouse.archives.gov/reports/katrina-lessons-learned/chapter1.html

*   *   *

## CONTRIBUTOR NOTES

**James A. Whelan, Ph.D.,** has taught chemistry at Lusher Charter High School in New Orleans since the school reopened after the devastation of Hurricane Katrina. A National Board Certified Teacher, he has spent 26 years teaching in Louisiana schools. He worked as an informal science educator and museum director for 15 years.

### WHAT'S NEXT?

In Chapter 7, you will read firsthand accounts of the Columbine tragedy and its consequences from the perspective of three who share what they learned about resuming educational services in the aftermath of a school shooting—the principal, a teacher, and a student. Unlike the disasters in the preceding chapters, the initial tragedy left infrastructure and resources relatively intact. Even though physically confined to the school, its effects reached out across the state and nation. As you will see in the following chapters, all tragedies, regardless of scale, bring long-lasting consequences.

CHAPTER 7

# VOICES OF COLUMBINE

CAROLYN LUNSFORD MEARS

TEACHING IS A CARING PROFESSION. EDUCATORS DEDICATE THEIR LIFE to the service of their students and, by extension, their community. Their goal is to guide students in the mastery of curricular content and, at the same time, nurture and reinforce attitudinal strengths that students will need for a fully productive and satisfying adulthood. Teachers sometimes endure sleepless nights, worrying about how to teach a complex concept so all students will understand, or how to reach that individual student who doesn't put forth sufficient effort, or what to do about a child who is withdrawing from peers and may need psychological testing. Yes, the questions that interrupt an educator's peace of mind are many, but thankfully, teacher preparation programs and professional development courses provide strategies and guidance for matters that fall within the norm of a school environment.

The world of educators, for the most part, operates within a certain range of predictability. Bells ring, class begins, Johnny's dog ate his homework, some students excel while others do not, and bus duty in a cold, drizzly rain is unbearable. However, on occasion, an event will strike that so fractures this world of predictability as to challenge the school's very existence. Few of us can imagine the experience of being a teacher in a school that is torn apart by gunfire.

On April 20, 1999, the business of schooling at Columbine was interrupted when two students, heavily armed with semiautomatic weapons and explosive devices, entered the building intent on causing maximum destruction and death. Having planned the assault for over a year, they set out, literally, to destroy the school and everyone in it. They sought to achieve a sort of dark immortality by becoming the *stuff* of nightmares for survivors, school families, the community, and the world at large. In less than an hour, they murdered teacher Dave Sanders and twelve students—Cassie Bernall, Steven Curnow, Corey DePooter, Kelly Fleming, Matthew Kechter, Daniel Mauser, Daniel Rohrbough, Rachel Scott, Isaiah Shoels, John Tomlin, Lauren Townsend, and Kyle Velasquez. They seriously injured 24 students and left countless others suffering with intense psychological and emotional wounds brought on by their experience. The tragedy of Columbine only began on April 20; its echoes continue even to this day.

The following accounts reveal the *inside* of a school tragedy from three perspectives: the principal, a teacher, and a student. It is based on my interviews with Frank DeAngelis, Beverly Williams, and Crystal Woodman Miller.

\* \* \*

### "YOU DON'T LEARN THESE THINGS IN PRINCIPAL SCHOOL"

*Frank DeAngelis*

Reclaiming Columbine High School after the tragedy seemed an unachievable goal. The world of Columbine had been shattered; the assumed future was gone. The job of leading the grieving school through the aftermath and returning it to its previous level of academic excellence fell on the shoulders of Principal Frank DeAngelis.

In the early days after the shootings, a community numbed by grief, horror, and disbelief stumbled to convocations, funerals, and memorial services. However, in spite of the psychological and physical turmoil, the school year needed to be completed. It was hoped that a return to classes could provide some semblance of routine, but first it was necessary to find a place to finish out the year, since the building had suffered significant physical damage and required extensive repairs.

Jefferson County School District considered alternatives, and it was decided that students would finish the year in split-sessions at nearby Chatfield High School. Schedules were revised, and Chatfield staff and students took the early morning shift so Columbine students and faculty could use the building in the afternoon. Principal DeAngelis describes the challenge of attending to everyone's needs in this changed environment:

> When we were at Chatfield, we had students and teachers that really felt the need to continually talk about where they were and what they were feeling. We had other teachers who wanted to go back to work that first day and just resume teaching where they left off on April 20th. And we had people in between. So meeting the needs of everybody was very, very tough. One thing I did, right from the start, was to tell people that we were all in different places, and we

may agree or disagree. That's okay, but if you disagree, we just need to respect each other for where we are. With any tragedy, whether it's in the immediate aftermath or even five or ten years out, people deal with it differently.

Counselors were available to help with the trauma, and students were encouraged to write out their thoughts and feelings. Boxes of journals were donated to the school for this purpose, and some students seemed to find relief in self-expression. Others felt increased stress and anxiety when asked to write about something that they could not find words to describe. Some students were being asked to write in their journals in all of their classes, with every teacher wanting to talk about it. Some felt inundated with "Let's journal. Where are you? How are you feeling?" It served as a constant reminder, and they were exhausted by it.

Columbine teachers, who were themselves traumatized by the attack, needed to prepare students for graduation just a few weeks away. However, their building was still in the hands of law enforcement as a crime scene, and there was no way to access student records to determine who had earned a diploma. Evaluating students in order to award final grades was also clearly impossible. Teachers did not have their grade books and course materials. According to school counselors, any senior who could walk across the stage that year was allowed to do so. Records were reviewed later so actual diplomas could be awarded to those who had earned them.

DeAngelis worries most about the graduating class of 1999. Many of them never returned to the school they had fled in terror. In the fall, they went off to college or the military or set out on another course, but they found they could not leave Columbine behind. At any point, the tragedy might be brought up, and they were susceptible to "meltdowns" without the support that was available to those still within the community. Some dropped out or returned home after first semester, unable to cope with trauma alone. DeAngelis still worries about what happened to them. "I'd be willing to put my neck on the line about this, but I think that alcohol and drug abuse and other risk behaviors went up greatly. These kids were saying, 'Don't tell me I've got a bright future when I just saw some of my friends killed.'"

RETURNING TO COLUMBINE

By the following fall, repairs to Columbine were complete so students and faculty could return to their own school. An upbeat *We Are Columbine* rally to reclaim the building began the school year on a positive note. While some grieving parents of slain students felt the ceremony should have had a less triumphant feeling, the large majority of students, faculty, and parents felt that after a summer of memorials honoring those who were killed, it was time to help the survivors "take back" their school.

No one expected the process to be easy. Principal DeAngelis reflects on his own experience in returning to the building:

> I came back before anyone else. It was tough. Each day I would walk out of my office and look down the hall. Each day as I walked out, I'd break down. I'd feel nauseous because of what had happened. I could visualize the gunmen firing and the glass breaking. Each time it would change a little, and I'd gain a little more strength. I talked to my counselor, and he said that all of us have a recorder in our minds, something like a DVR, and we can decide what we want to play on that recorder. We can decide if we want to erase the old scene, if we want to fast-forward it, or if we want to rewind it, but the important thing is that we can be in control. So, that's what I would do when I'd walk out of my office. Even now, when I'm in this building alone by myself, I still kind of tingle and get chills. But I can choose to be the director of that movie, and when I walk out, I no longer envision the gunmen. Instead, I envision students laughing and joking on a normal school day.
>
> It's the same thing for the 13 who were killed. For years I struggled with thinking of Cassie dying in the library, Isaiah dying in the library, and Rachel, and the others. When I went from seeing them being victims who were killed, I was able to focus on their life. I'd imagine Rachel Scott starring in the school play that year, and I'd visualize Lauren Townsend playing volleyball, and all of the others. Playing that movie differently really helped me.

DeAngelis credits a friend, a Vietnam veteran, who called him within 24 hours after the shooting and advised him to take care of himself *first* so he would be able to take care of others. His friend told him what he wished he had known about trauma when he came home from the war, but didn't. DeAngelis sought counseling right after the tragedy and still "goes in for maintenance" whenever he has recurring dreams or realizes his attention span is not what it should be. He has been in five or six minor traffic accidents, "just fender benders," because his attention wandered. These lapses in concentration usually happen about the same time of year—in April, the time of the tragedy, or in October, around Dave Sanders's birthday. DeAngelis has concluded that "the mind does funny things, but people don't know what to expect from a tragedy until they're experiencing it. That's unfortunate. It would help to understand it before you need it."

Students, teachers, staff, and others connected to Columbine faced many of the same challenges. Since the assault had been planned and carried out by "two of our own," they faced additional anxiety, a deep sense of betrayal, and loss of trust—in classmates, in adults, and even in themselves, for no one had considered the gunmen as real threats prior to their rampage. All wondered, who could think that someone sitting in their class would be plotting their murder?

Returning to school meant facing these demons every day. Teachers and staff had the option of transferring to another school, but almost all chose to stay at Columbine and were granted extra time off for personal leave. Students were also granted transfers upon request, and any who could not return to a classroom setting were provided with home-based tutoring. To minimize potential for re-triggering the trauma, mental health counselors advised changes to the physical environment (e.g., new paint and flooring) in addition to the essential repairs. Advance notice was given whenever a fire drill was scheduled, and the sound of the alarm was changed. Columbine became a "balloon-free school" for years, since the sound of a popping balloon triggered traumatic memories of gunfire in the halls.

The first year after the tragedy was especially difficult for anyone new to the school. An attitude of, "You weren't here—you don't understand," developed and that created a very difficult situation for the incoming freshmen class and for new teachers and staff. If they saw someone crying and they cried too, someone would say, "Well, what are you crying about? You weren't here—you have no idea what I'm feeling." It was impossible to accommodate everyone, and DeAngelis often said that if he could satisfy 70 percent of the people he worked with during the day, then he counted it as a successful day.

After the tragedy, all were "walking on eggshells." While some needed to talk about it, others were saying, "We are tired of talking about it and we can never move forward until we stop talking about it." But it was an issue that couldn't be ignored, and from one day to the next, the situation might change. People who were *up* one day might be *down* the next. Other troubling events followed: the release of police reports, lawsuits, Internet threats to the school, suicides, and murders of two Columbine students at a local sandwich shop—all things that had people on edge. "Not a day went by within that year in which there wasn't something written or spoken about Columbine in the media."

## RETHINKING CURRICULUM AND INSTRUCTION

Columbine teachers reconsidered their instructional plans, carefully screening all activities and texts they had intended to use for their classes. They assessed how the activities they had planned might affect their students. Videos that were going to be shown for social studies class, for example on World War I or World War II, were eliminated because of the guns and the sounds of gunfire. Novels that described any violence were dropped from the curriculum. It became clear that people could be re-traumatized by anything that involved the senses—smell, touch, sound, sight. For some it was even taste, like the food served in the cafeteria that day. For students trapped in a storage room with cleaning supplies, it was the smell of cleanser. The list of possible triggers seemed unending.

In the year after the tragedy, academic performance was understandably lower than in the past. Principal DeAngelis understood; he could not even read a newspaper—just couldn't concentrate: "How could you learn in a situation like this? How could you concentrate on physics and chemistry when you had had this experience? How could you learn when you were worried about your safety?"

"I can't imagine what it would have been like to try to do complex math. Post-traumatic stress is going to affect you." In the spring before the tragedy, DeAngelis had planned to enter a doctoral program but after the tragedy decided not to, knowing that he lacked the necessary concentration and that his priorities had changed, "so I know it had to have affected the kids."

Attending to the instructional needs of students when everyone was in the midst of trauma was a major challenge. "You could have the best aligned curriculum and instruction plans in the world, but they're not any good if no one in the school feels safe, so that becomes the priority." Denying that the tragedy had happened and trying to return to the way things had been, of course, was impossible, but some felt that if you didn't talk about it, it would go away. Whenever DeAngelis had an opportunity to share his thoughts with students, he would use himself as an example:

> I would stand up in assemblies and ask, "How many of you people are scared?" No one would raise their hands. And I would raise my hand and say, "I'm scared to death every day. And it's not going to go away." My modeling let students and faculty members know that they weren't alone. Often it was that support system that got us through.

Addressing calls for increased security required reasoned judgment. Some parents and community members felt that armed guards and metal detectors were needed. People came in offering all sorts of security programs, but the students were telling their principal, "Mr. D., I don't know if I want to stay here because this doesn't feel like a school—it's a fortress." A fine line needed to be recognized. Reclaiming the school meant reclaiming it as a place of teaching and learning instead of allowing it to become an armed camp.

Principal DeAngelis acknowledges that in the aftermath, his decisions were more likely to be on the side of more discipline. Students made a T-shirt that said, "If I offend, they suspend." There was a delicate balance that he and the faculty were working toward:

> If the kid makes a comment, do we assume that there is another murderer out there? At the same time, the entire school was constantly under the microscope. Everybody was on edge. Every action, every decision, was being

examined by millions because it was reported in the media, so if you didn't do something, then people would say, "Those people at Columbine didn't learn. Didn't those 13 lives mean anything?" But at the same time, I might have parents coming and asking, "If this wasn't Columbine, and that hadn't happened, would you treat this situation the same way?"

Some people approached safety issues after the shootings by advocating changes in school district rules and policies. For example, questions were raised about dress codes because the gunmen wore black duster coats. Some insisted that Columbine needed to adopt a *zero tolerance policy*, but as DeAngelis pointed out, Columbine had always had such a policy; it just had not been called that.

> Other people gave it that name and said that now we would have to start disciplining kids for harassment and bullying. But we had that policy and did that before. People were acting like Columbine was a free-for-all school where kids could do anything they wanted, and teachers would turn the other way. But it wasn't like that.

In some schools that have had a shooting, district officials have reassigned the entire administrative team, but at Columbine the team was kept intact and any who wanted to return could do so. This continuity provided a measure of support for the students and staff, and DeAngelis promised that he would be there for them:

> Whenever I talked to students, I would tell them, "I hope you will return to Columbine, but if you don't, I support you in your decision. And when you are ready to come back, I will be here waiting." I could not stand there and ask them to continue with their education at Columbine if I wasn't going to be here with them. I made the commitment to stay at least through the class of 2002, when the last of the students who had been here during the tragedy would graduate. Now, I've made the commitment to stay until the class of 2012 graduates. By then every kid who was in the Columbine articulation area in April 1999 will have graduated.

## REDEFINING "NORMAL"

After a traumatic event of any kind, the deepest desire is to return to some sense of normalcy. Life has suddenly become unpredictable and threatening. Well-intentioned people reach out to help and may try to reassure survivors that "time will heal," as if one day it will all go away.

> Many people expected there to be some magical day when they would wake up—maybe the one-year anniversary or something like that—and everything

would be fine. But that's not what happens. As it turned out, many people were doing okay, and then the 10-year anniversary comes up and they'd start experiencing things they had not experienced before. You can't predict what's going to happen and that's why it would drive me crazy; people saying, *get over it, get over it, get over it.* But it doesn't happen.

The one question people ask that bothered me then and still bothers me today is, *Are you back to normal?* I say this time and time again. We will *never* be back to normal. That old normal doesn't exist anymore, so you have to just redefine what normal is. It's a different normal for Columbine High School.

In the wake of the shootings, with everyone working toward a new normalcy, each day presented a different challenge, and the effects of the tragedy were widespread. Parents and siblings of students, spouses and children of faculty and staff, even members of the community with no direct connection to the school struggled to cope. There were some "very messy divorces," elevated interpersonal conflict, domestic violence, increased alcohol consumption, and suicides. Parents were at a loss for how to help their children, and this uncertainty found its way back into the school. When parents asked the principal for advice about what to do for their children, he encouraged them to take them to counseling, whether they wanted to go or not. Not everyone took that advice: "Parents have come back, even 10 years later, to say that they wish they had listened because now they are dealing with issue after issue after issue." DeAngelis compares the situation to a person with a drug or alcohol problem who doesn't want to get help because it's more comfortable to stay in denial. "There comes a time, as a loved one, that you need to intervene and get them help because they're not thinking clearly. It's tough love at times. And that's hard for parents to do."

Scars that come from traumatic experience, whether they are physical or emotional or spiritual, are deep and long-lasting. Some people are affected instantly, but for others it won't show up for years. At Columbine, some tried to "stuff it" and just move on. A victims' assistance program was available for three years, but when it ended, some were just beginning to need help.

Unfortunately when things like this happen, it's like what occurs when there's a death in the family. All of a sudden you have all of these support systems. You have relatives, neighbors, and friends who come to give you support. Where people are really going to need the help is down the road, after the family has left town or the neighbors have gotten back into the routine of their own life. It's down the road that people will need them—need their support. You have to plan ahead and realize it can happen and know where people can go to get help.

Another point that DeAngelis makes about leading a school after a tragedy is that sometimes people will take advantage of the situation. He offered the following story as an example:

> We had a suicide of a young man about five years after the tragedy. We went around to each of his classes, and we would talk to the students in that particular class and tell them that if they needed to, they could go down to the office and talk to the counselors. If I had to do it all over again, I would do it differently because what happened was every kid went down. Every kid was his best friend. So I said, "Ok, here we go—let's see what happens now."
>
> As it turned out, most of the kids were legit, and once they got some help, they went back to class. But there were some girls who came down here the entire time, and I called them to my office and told them, "You know, I'm really concerned about your well-being. You're having such a hard time, so we'll call your parents and send you home for the next few days. We'll have the counselors call you." Within five minutes they were out the door and back to class. I think anytime you have a tragedy you're going to have some of that.

## LEADING FROM THE HEART

Frank DeAngelis describes his leadership style as "leading from the heart," and notes that the skills he used on the day of the tragedy and afterward were those he had used throughout his career. He believes his reputation for being fair, compassionate, and consistent had helped establish his credibility, so everyone knew what to expect from him.

> The Friday before the shootings, at the before-prom assembly, I told the kids how valuable they were and that I wanted to see every one of their smiling faces back at school the following Monday. I had them close their eyes and visualize what it would be like to have the person sitting next to them *not* be there because of making unwise choices. I told them I loved them. So then just a couple of days later, the shootings happened. When I saw them the next day, it was just me being me only in a different situation. I told them how I felt. Counselors would come in and say, "Frank, by you being up there and showing emotions, you are granting everyone else permission."

For Columbine, perhaps the most inspirational example of compassion came in the person of Gerda Weissman Klein, a Holocaust survivor who reached out to the school and offered her heartfelt support. She has visited the school numerous times and is still in contact with the Columbine community. Many are inspired by her powerful message: "Pain should never be wasted." Whether chatting with individual students or addressing whole school assemblies, she inspired many associated with Columbine to transform their own pain into a helping spirit. DeAngelis declares that she is probably one of the biggest influences for him:

> Sometimes people come to help, and they're very knowledgeable but in the back of your mind, you're thinking, "How can you know what I'm feeling— you haven't lived this?" But with Gerda, understanding was naturally there because she had survived far worse than what I did. Here's this person who was 17 years old and in a death camp, lost her whole family, yet she survived and now she comes to help us—at the age of 85, she's still going strong.

Like Gerda Klein, Frank DeAngelis reaches out to others who may be hurting. He contacts administrators of other schools that experience a shooting, offering his support and encouragement as one who has lived through it. It's an unfortunate "club" to become a member of, but this school leader knows that, as someone who understands, he can help.

> Sometimes people want to ask me questions about Columbine but are afraid to bring it up. It's like they are walking on eggshells, and they tell me that they are afraid to ask me anything. I tell them not to be afraid to ask about it. Helping other people is therapeutic. The only times I will share my story, is if it will help someone. That's the only thing I can do.

## ACCEPTING THAT CRISIS IS POSSIBLE

Even though people realize that school shootings happen, many believe themselves immune to such an event. It seems human nature to ignore risk, when such occurrences are so far outside the range of normal experience. DeAngelis admits that he too felt this way:

> If you had run a scenario of a "Columbine" shooting happening at Columbine, I would have said, "It just won't happen." That was a lesson I learned. Unless people are directly affected by a situation, they aren't going to think it could happen to them. They won't consider the possibility. But when you're actually faced with a crisis, you're under so much stress, that it is hard to learn or to listen to anyone. You need to think about it in advance, but people don't want to do that.

Leading a school through a crisis and its aftermath requires more than awareness of potentiality. Taking the threat seriously means taking action, and that includes not only planning and practicing, but also assessing how people might perform in an actual emergency.

A school can have the best laid out crisis plans in place, and you practice, practice, practice. But you cannot predict exactly how people are going to react under fire. As a leader, you hope that you know your people well enough and that you have people in positions that can handle the situation. During the crisis, if you have people in key positions that react irrationally, then it will just add fuel to the fire.

Even with the best plan in place and responsible personnel in key positions, predicting behavior in a real emergency is difficult. DeAngelis adds that there is always the element of the unknown:

> You hope that people will react well in a real crisis, and sometimes that may mean *not* following the plan because many times what is planned does not address the situation you actually face. People have to make those split-second decisions, and you cannot predict what they will decide. Think about people who were on the planes on 9/11. What caused certain passengers to fight the terrorists while others sat back? What allowed that to happen?

Preparation for an emergency also requires that a leader assess his or her own potential response. A crisis situation does not allow time for self-reflection.

> One time, we were doing our regular emergency response training, and usually the principal is what they call the district's incident commander. I told the sheriff and the head of school district security office that in the shooting situation that happened here, I probably violated every plan we had, because as incident commander I should have gone to the command post and stayed there. But on April 20, my first instinct was to go out and help the kids and the staff. Hopefully, it never happens again, but if it does I'm doing the same thing. They told me that if my people know that, and I've designated someone ahead of time to be incident commander, then that would be okay. I think it's important to think about things in advance. You need to know in your mind how to play it out. Think about how *you* are most likely to react and plan according to that.

Dr. David Benke, a math teacher at nearby Deer Creek Middle School, exemplifies this type of thinking. When a mentally ill gunman opened fire on students outside of Deer Creek, Benke tackled the shooter and held him until police could arrive. DeAngelis credits the teacher for having played the scenario through in his mind for months.

> He had attended emergency preparedness training and had thought about it enough to know what he would need to do. He told the kids, "If a gunman

ever comes here, this is what I'm going to do." So when it actually happened, he knew what he was going to do. There was no second guessing, no second thoughts about it. And he saved lives.

It is important for leaders to recognize that people need permission to make decisions on the spot. While some may take action to confront a threat, others will respond differently. At Columbine, there were teachers and staff who helped students escape when that was possible and sheltered them in classrooms and closets when it wasn't.

> People did the best they could. They saved lives, yet there was a lot of sur-vivor's guilt. It's a tough thing. There were so many unsung heroes that day. Dave Sanders lost his life, and there were others who put their life on the line for those kids too. Theresa Miller, the chemistry teacher, extinguished a pipe bomb that went off. The custodians kept going out and shepherding kids to safety. Other staff members were doing the same thing. But there were several who didn't come out because dispatchers told them they needed to stay in place. Those who were outside of the building weren't allowed to go back in and help. It was tough.

While a crisis in a school presents a common threat, each individual will have a different experience and a different response. In reclaiming Columbine, DeAngelis made it a point to respect each person's actions and decisions, acknowledging that lack of information during the extended event made the situation so much worse. In the years since the tragedy there, communi-cation systems have been improved and protocols for tactical response to a school under siege have been rewritten.

Lessons that were learned from the Columbine tragedy have helped save lives at other school shootings. And DeAngelis's model for school leadership, leading from the heart, inspires and educates those who may face similar tragedy in their own schools.

\* \* \*

## "TAKING CARE OF THE CHILDREN"

*Bev Williams*

GUNSHOTS IN THE BUILDING

April 20th was like any other spring day at Columbine High School. The energy level of the students was high as seniors were getting ready for gradu-ation. My fifth hour class began at 11:15 a.m., and on the agenda was a quiz on nutrition and digestion. However, as a follow-up to "Diversity Day," two weeks earlier, I asked one of my students to give us a brief demonstration of break dancing since he was known as the "master of the dance." His background music must have obscured the initial sound of gunfire from the parking lot because within a minute of completing his performance, as students prepared to begin their quiz, we heard a huge stampede coming down the hall.

My classroom is located at the top of the steps leading up from the cafeteria and approximately 40 feet from the school library. One of the students asked if he could see what was going on. Most of us assumed that a senior prank was taking place. I agreed but asked him to do it quickly since the quiz was about to begin. As he opened the door, we could hear popping sounds that we all assumed were firecrackers. Outside the door were two colleagues, the chemistry teacher on hall duty and our technology specialist. Color had drained from their faces, and they shouted to get all the kids down because there were gunshots in the building.

Several students who were racing down the hallway were shoved into my room, and realizing that we were extremely exposed in this corner classroom, I directed the kids to stay down and to hurry into the "greenhouse." The greenhouse, a small storage room that adjoins my classroom, earned its name because we keep a few plants there, plus it contains a sink and drain. One of the boys was paralyzed with fear and had to be coaxed to move out of the classroom.

I honestly have no recollection of what happened next until I realized that I was sitting on the floor in the greenhouse with my back leaning against the closed door, and 31 pairs of frightened eyes looking at me. There are two doors into the greenhouse, both having windows that I had fortunately covered with posters several months prior.

I was 85 percent certain that the door between the greenhouse and the main hallway was locked, but the door adjoining my classroom, against which I was sitting, could not be locked from the inside. I felt that if we remained quiet, though, whoever was orchestrating the siege would not know we were there and we would be safe. Of course at this time we had no idea how many gunmen were involved.

Moments after we were settled, I became aware of a piercing fire alarm that had been activated. It continued to sound for hours. Because of our proximity to the main hall, the library, and the cafeteria where so many of the bombs were detonated, we heard a lot—single gunshots and semiautomatic fire as well as bomb explosions.

We were extremely cramped. Many of the kids were trembling uncontrollably from fear; several were sobbing softly; and all faces were white with terror. We were in the eye of a hurricane, and yet had no real idea of the danger we were in. It was the quintessential scenario of "ignorance is bliss."

Even though all classrooms had phones and a TV monitor, the greenhouse did not, so we had no communication with the outside world. We didn't know if there was anyone else left in the science department or for that matter, in the school. We sat holding hands, softly rubbing one another's back to stay calm, and silently praying. When I first looked at my watch in the greenhouse, it was 11:30.

The sounds of gunfire and bomb explosions continued sporadically throughout the ordeal. Fortunately, the continuous ringing of the fire alarm helped to muffle much of the sound and distorted the distance and direction from which it was coming. About 20 minutes after it all began, we heard one of the gunmen standing outside the door to the hallway scream with a taunting, bravado voice, "Today I am going to die!"

We often heard the gunmen talking to each other but could not decipher what they were saying due to the background noise. We did not know it at the time, fortunately, but my classroom door was unlocked and they did enter my room. They could have so easily tossed a Molotov cocktail in, broken a window out, or done whatever they chose. They did that in some rooms. I think that when they saw my room empty with the overhead projector light still on and backpacks lying everywhere, the shooters probably assumed that everyone had left the science wing so they did not look closely in the rest of the rooms where others were hiding. In retrospect, the number of fatalities could have been so much worse; there is no doubt in my mind that divine intervention occurred. The more stories I heard from faculty and students, the more strongly convinced I have become of this fact.

We sat quietly in the greenhouse for almost 3 and a half hours. It quickly became apparent that the kids in all their innocence expected me to know what to do. Often I thought to myself that I wasn't prepared for this. Kids asked me questions for which I could not possibly have an answer (Have the police arrived? Did my sister make it out unharmed?). Since I was a teacher, they assumed that I knew all. I felt like I was a shepherd with all of these little sheep I needed to take care of, so I told them what they wanted to hear, just so we could get through it. What's interesting is that no one ever questioned my responses, not realizing that I was no more privy to information than they were.

At one point, when we didn't hear any more for a while, I was going to open my door and just see where the other teachers were. One of the kids said, "Please don't leave me—please stay here." I said, okay. At some point I knew the police would get us out, and until then we just would sit tight.

We saw the first news helicopters shortly after the siege began and continued to see them as well as small planes circling the campus as the events progressed. But we still had no idea of the scope of the situation. The height of innocence was epitomized when over 3 hours after the tragedy began, one of my sophomores asked if I thought we might be on television that night. Wow, what an understatement! Time passed so quickly and yet so slowly. The room frequently shook from the explosions and gunfire. Periodically the kids had to stand just to keep their circulation going. Our arms and legs occasionally lost feeling due to the cramped conditions. I can recall thinking that if our men in Vietnam could withstand being in foxholes for long periods of time, and if Holocaust survivors could make it through years of terror rather than a few hours, then we could too.

Occasional moments of levity offered comic relief. One of the girls wedged her head between a helium gas cylinder and a cabinet, which distorted her face as she tried to extricate herself. It gave all of us a welcomed giggle. And of course the obvious biological problem arose when nature called with nowhere to "go." A couple of the girls took care of necessities in the sink while the boys turned their heads, and the boys took advantage of a bucket we affectionately referred to as the "Pee Pail." The first boy to use it was so scared that amidst relative silence in the room, a few pathetic drops echoed. One of the other boys whispered, "Come on, you can do better than that," upon which the floodgates opened. No one could conceal their laughter.

Time dragged on. After about 2 hours, the fire alarm ceased, and the all-clear signal began, but rather than shutting off after 2 or 3 minutes, it continued to sound. All of the class bells continued to ring on schedule, and when the 2:30 bell sounded indicating that school was over, one of the girls whispered softly, "Oh, good, now we can go home."

## ESCAPE AT LAST

The kids were getting impatient after hours of sitting, but I kept explaining to them that even though we knew we were safe, their parents were experiencing the worst anguish of their lives not knowing their fate and that each one had to contact mom or dad just as soon as we were free.

After 2:30 p.m., things were rather quiet and the kids asked when we could leave. I told them that we would not leave until we heard from a police officer or other adult that it was safe to evacuate. That welcome time occurred at 2:50 when a loud male voice coming from the hall outside our door shouted for his men, "Raise your shields!"

The kids were leery about opening the door but eventually they did. The SWAT team wasted no time in getting us on our feet and frisked. We were instructed to keep our hands on our head and to run across the hall and down the steps to the first landing because we were in the "line of fire." We did as we were told and were horrified to run across a rug that was bloodstained.

On the landing of the stairs we were again frisked, and as I looked into the cafeteria, I could see that many of the windows had been blown out, broken glass was everywhere, uneaten lunches were left on the trays, backpacks lay abandoned all over the floor, and the sprinkler system was fully engaged. It resembled a war zone.

With our hands on top of our head, we sloshed through four inches of water and were led across the back of the cafeteria and out a side entrance only to encounter the bullet-riddled body of a young man on the sidewalk. We ran up the outside steps, only to see a young woman lying in the grass in the same condition.

Once behind the school we were again frisked and pushed into SUVs. Grabbing on to one another and with heads down, we were whisked cross-country over sidewalks and bushes to an area in a nearby park where an Emergency Center had been assembled. Only after I stepped out of the vehicle and saw the hundreds of rescue workers, helicopters, ambulances, bomb squads, firemen, police officers, and multiple news crews did I have any idea of the magnitude of what we had experienced. We were quickly examined and questioned by several teams of investigators. The kids were physically led to cell phones so they could contact their parents while other kids walked by aimlessly in a daze. I will never forget the scene.

Once my students were cleared and on their way to a nearby elementary school to reunite with their parents, I started looking for other teachers and colleagues yet could find none. Eventually I saw more people from the science department being evacuated and realized that we all had survived. It wasn't until I crossed the cordoned-off area and entered a large parking lot that I could see about ten teachers who had made it out during the early moments of the ordeal come rushing towards me with their arms out screaming, "Bev, you made it, you made it!" Emotions ran high.

MAKING SENSE OF IT

I cannot begin to understand how my family and the parents of the children who were trapped with me suffered during this time of such uncertainty and helplessness. And, yet, how much more devastating was the grief for the parents and siblings who learned so much later that they had lost their loved one. What a terrible, terrible, tragedy!

I am still puzzled as to why this particular school shooting touched the heart of the nation in such a profound way. Was it the scale of the tragedy that made us sit up and take notice? Was it the fact that the nation watched in horror for hours while the media broadcast live coverage of a real life drama without a very quick or very happy ending? I personally think that Columbine represented the American Dream, and now that has been shattered in a very real way. I recall hearing Tom Brokaw and Katie Couric report that they had had a chance to tour the Columbine community and likened it to the "Brady Bunch" community or a "Leave It to Beaver" neighborhood. Is this why it affected people so deeply? I guess we all wish there were simple answers, but unfortunately I don't think they exist.

I discovered that "survivor's guilt" is very real and all of us experienced it to some degree. Probably the person in our department that suffered most deeply was this wonderful teacher who came out of his classroom and saw what was going on. Another teacher was standing there and told him to run down to the main office and tell the administration what was happening. So this teacher ran down the hall and as he got near the front of the building he could hear bullets going past him, breaking some of the windows in the office. He made it to the office and tried to find Frank—our principal. He led about 12–15 counselors, assistant principals, and others to safety, but when he tried to come back to help others, he wasn't allowed to. He went all around the building trying to find a way in, but he was not allowed to return to his room, to his students. He was stranded outside the building having unimaginable concern for the rest of his colleagues and the students who were trapped inside. That was so very difficult for him—not being able to help. It was heartbreaking.

I also learned that all of us grieve very differently and at a different pace. I think that those of us who witnessed the tragedy will forever alter our priorities. The event itself was so surreal that it was actually the issues and events that occurred in the months afterward that were the hardest to deal with, such as . . . the media.

The media—where do I begin? There were thousands of media personnel everywhere we turned. Many were extremely sensitive, but some were not. I honestly never thought I would watch a CNN or national news broadcast or open a *TIME* magazine or *Newsweek* and not only know all of the people on the cover, but also all of those being interviewed. I never thought I would turn down an interview with the *New York Times*. Our school district asked teachers in the science department to give an interview to *People* magazine since the reporter had children in one of our schools. We agreed to do it, as long as a district representative was present, so if any legal concerns needed to be addressed, he would intervene.

The outpouring of love was tremendous. In the last few weeks of school when we were on split-session with Chatfield High, 20 volunteers were reading an average of 20,000 letters a day from across the nation and world. People in Texas sent a teddy bear for each one of our kids, and it was awesome to watch our big "jocks" clutching their bear just as much as everyone else. Millions of children sent cards and banners; Mormon women made each of us a quilt for comfort; thousands of school supplies were donated by

Walmart, Office Depot, and so on. The Rockies organization put "CHS" on each ball player's sleeve as well as in a heart behind home plate at Coors Field. They provided us with complimentary tickets; local restaurants gave us complimentary meals. The Pope even asked the Cohen brothers who sang the nationally broadcast song *Columbine, Friend of Mine* for an autographed copy and a private audience. The visits by President and Mrs. Clinton were amazing. Even the most cynical of teachers and students were awed by the Clintons' sincere expression of sorrow and the president's coming to visit with them later and his unwillingness to leave until every student, teacher, or rescue worker had a chance to shake his hand or give a hug if they so chose.

I will be forever grateful to the police and SWAT teams for risking their lives to save ours. It's amazing what great wisdom hindsight can provide, and I become so frustrated when I hear criticism of their efforts. Given the little information they had at the time and the quick decisions they had to make when entering a huge school full of trapped children, with bombs exploding, and unaware of the number of gunmen or the type of fire power they would encounter, I felt they did an amazing job. In so many ways, the Columbine story is truly a story of heroes.

Right after the tragedy, one of the young teachers in our department looked at me with so much sadness and said, "Bev, I entered teaching to influence lives—not to have to save them." I think that this pretty much says it all.

## THE AFTERMATH—NOTHING NORMAL ABOUT IT

The kids were very fragile and we knew that we had to provide a very safe environment, no matter what else we did. We also needed to be as normal as possible in a situation that wasn't at all normal. When we returned to classes at Chatfield, the very first time I saw them they could not get into my classroom without giving me a hug. I think we all needed extra hugs back then.

We all tried to make accommodations for their learning needs. Being able to bend was so important. We had to continually adjust. One thing I love about teaching is that kids have so many different learning styles, and I like to find activities so that they can learn even when they have different ways of learning. Yet, there are some teachers who are not comfortable with modifying instruction to different learning styles. They just want to find something and they would use it the same way for 20 years. It was much harder for them after the shootings.

In the science department, we developed activities that allowed students to work in groups. That way they could interact with their peers and not just sit by and disappear into their own silences while listening to the teacher.

Even the final exam was designed as a group activity. None of those assignments or the final counted of course, but the students didn't know that because that would have been outside of the realm of what is normal, and everyone needed to feel some sense of normalcy. So we planned activities that looked like school, even though they were not *real* school as it had been before.

I had my students plant flowers—columbines—that they could take home and transplant in their yard. It gave them a chance to be in a position of nurturing something and watching it grow. At the end of the year, some students took them home, some did not, but it just gave them a chance to do hands-on work, and always with others. On occasion we would plan activities to help with relaxation—one on humor, laughter, was especially good.

Another thing we did in a group setting was to create concept maps, a visual representation of how all of the major concepts they had studied were related. So many times students see what they are studying as isolated facts, but with concept mapping they could put things together and see how one concept is related to others. In making the maps, it allowed making connections but it was also sensory, because they were drawing and using colored pencils to show relationships. Again they worked in groups of three or four, and it really helped the kids—it was not a demanding activity, but they had to understand the concept in order to demonstrate the relationship.

The next year, we were still just trying to make it through, and we let students know that no matter what fears they had, they were going to be okay. We tried to keep things as normal as possible, though of course there was still absolutely nothing normal about it. In some ways, I think I tried too hard to have the same expectations that I had before. I don't know if that was really the right thing to do or not. For some kids it was comforting because they could really get into their school work and focus on that, but for others, I'm not sure it was the best way. It may have just added to their stress.

None of us knew what to do. We wanted to do what was best and didn't have a handbook on how to do it. We were always very responsive to what the kids needed. I have to admit I never thought that when I went to a parent conference I would be talking about what type of antidepressants or medication the kids were on, and that the teachers were on. The academic performance was certainly on the back burner. I thought, boy, the whole focus of a parent conference had been changed.

We had no precedent for what we were doing, but we were there for the kids. I remember that the head of the school board stood up at a meeting and said, "Thank you, everyone, for saving so many of our children." That was the first time we had been thanked. It may sound like a small thing but it was huge for us because then it was clear, he understood. He understood what we had done. I think most of us had only done what ordinary teachers tend to do, which is take care of the kids, but it was important that he recognized that we had gone through something dreadful, but that could have been so much worse. Just that heartfelt and sincere *thank you* meant so much.

After the school had been released as a crime scene, some of us teachers asked if we could go into the library, but we were not allowed to do so. Columbine teachers were not allowed to go in, and yet others were allowed in. A community leader told me that he had never been into the school before, but he was allowed to go into the library after the shootings. However, I couldn't go in and it was *my* school. It felt like we were not respected as professionals, and that was very difficult. Teachers needed to be allowed to choose for themselves. Maybe they didn't want to go in, but those who did should not have been denied that opportunity.

There was a lot of controversy about what to do with the Columbine teachers after the tragedy. Were we to go back to the school, or was the district going to reassign us all to different schools? We felt at that time that we really needed to go back, to be with other people who understood us. In the long run, I'm not sure that was the best thing. For the students it *was* best, without a doubt, but I'm not sure about for the teachers. The new teachers who came into the building afterwards always felt like they were outsiders. Everyone tried, but I don't know if it would have been better to break us up or not.

## THE "GREENHOUSE CLUB"

When we were at Chatfield, one of the girls who had been in the greenhouse during the shootings said, "Mrs. Williams, you know we should start a club—a greenhouse club—and you could be the president." So, I asked her what office she would hold, and she said she thought she'd be the secretary. So we started a "club" for all of us who had been in the greenhouse on April 20. She made up an "official" Greenhouse Club certificate, and had a membership number on each one. I told the others that there would be a lot of people who would want to join us, but we were the only ones who could join—the charter members—and that was a distinctive honor.

It was sort of ridiculous but recognizing our connection—making some absurdity from something so awful—helped. It was incredibly powerful. We ended up having a group picture made the next year. That Christmas I sent everyone a little note and a copy of the club photo. Actually we had a picture made several times because I thought it was important for the kids to come together, so we would get together periodically. In the kids' senior year, we were going to get together and have another group picture made, but very few kids showed up for it. I realized that they didn't need it anymore.

In the years since the tragedy, students have said that being part of the "club" gave them a sense of community and belonging. It seemed to help for them to know that there were others who understood their experience.

## ONE FINAL REFLECTION ON *THAT* DAY

It was interesting for me to realize how differently the men and the women teachers responded to the situation that day. My colleagues all observed that during the shootings, when we were all pretty much trapped in the science department, every male teacher, without exception, picked up something to use as a weapon. One of them picked up a metal stand that's used to hold beakers. Another picked up a stapler; another grabbed a paper punch. Every male teacher picked up something and held it as a weapon. And the women—we were in that nurturing role, sitting and holding children, attempting to soothe and comfort. It is so interesting how very different from men that the women were. Some of these things are so primal; they are so engrained in our DNA that it cannot be dismissed. It was without exception.

I will also add that, since we were all trapped and didn't have another way out, the science teachers were all absolutely adamant that the district build a second exit from our area. That was something that had to be done before we could feel safe. I still look for exits when I walk in almost any building. I guess I always will.

*   *   *

## "NEVER, EVER BE THE SAME"

*Crystal Woodman Miller\**

It's lunchtime, and I have a physics test looming. I haven't studied, and I know I will have to pull off a serious cram session if I am to pass. The bell rings and I grab my backpack and fly to the door. I need to go study! In the hall, I am greeted by my friend. I grab his arm and drag him to the library to help me. Surely the library's the place for some quiet study time. On our way, we find my friend's sister and drag her along too.

The library is strangely empty as we walk in and take seats in the middle section. I consider the challenge of preparing for this test and sigh as I swear

\*This section includes several excerpts from *Marked for life: Choosing hope and discovering purpose after Earth-shattering tragedy*, by Crystal Woodman Miller with Ashley Wiersma, © 2006. Used by permission of NavPress. All rights reserved. www.NavPress.com

for the *n*th time to stay on top of my schoolwork. While I'm in mid-thought, a teacher bursts into the library. "There are guys with guns!" she screams. She is frantic, rushing back and forth across the east side of the library. "They've got bombs! Do you hear me? Bombs and guns! They are shooting at students! Move...now! Get under your tables *now!*" she belts. Her voice is cracking. She is as white as a sheet.

There is palpable tension as all of us stare at this hysterical woman—*Is this a joke?* I stay in my chair, waiting for everyone else to make a move that will tell me this isn't for real. But in the halls outside the library, I hear students running and screaming—it all sounds so distant, so bizarre.

A student stumbles into the library—he's bleeding! We crowd in beneath the table, not knowing what is happening. My friend tries to comfort me, but suddenly an explosion goes off in the hall.

With all my heart, I want to believe my friend. Nothing bad ever happens in Littleton, Colorado. This is our town—our secure, intact bubble of alrightness. There are no gang problems here. There's no huge drug ring in operation. I quickly focus on the other school shootings I'd heard about, deciding that Littleton was no West Paducah. It's *safe* here!

Angry sounds are heading straight for the library, rapidly getting closer. As I move closer to my friend, I hear two evil voices booming, in the library now, filled with spite and anger and rage. I hear the taunting, the gunfire, the laughter, and the celebration as students are shot—as students are dying.

I contemplate my own death for the very first time and see images of family and friends rush across my mind's eye. I am gripped by instant replays of the stupid decisions I've made, the ridiculous things I've done in my past. I start to wonder what it will be like—what it will *feel* like to be shot. I prayed, *God, if you are real, get me out of here, just get me out of here and I'll serve you forever.*

Gunfire sprays the library walls and several bombs explode. I choke back a cough from the smoke, hesitant to make a sound. They are coming toward our table. I am lost in despair, swimming against a tide of anger and confusion and fear and outrage and bitterness and resentment. A blast of gunfire, as they shoot a boy at the table next to ours.

"I dropped my clip," a shooter screams.
"Look for it!" the other one yells.
"It's too smoky, I can't see it," comes the response.

I hear the voices retreat, out of the library—they're going to reload. My friend yells at me, "We have to go—now!"

As if awakening from a horrifying coma, we run for the back exit—past the carnage, the crumpled bodies lying near tables and beside computer stations, lifeless and surreal.

I stumble back as the outside air hits my face. My eyes are fixed on the exit path. Kids are darting everywhere, some bloodied by gunshots and bomb shrapnel.

I am outside—standing below Mount Columbine—at least that's what we called it. I'd run that hill a thousand times for track drills, but that afternoon, it was as if I'd never been there before. Somehow, despite the masses of people around me, I felt completely and utterly alone. Even in my state of shock and disorientation, I remember thinking how odd and eerie and cold the silence was. The sky turned gray and gloomy as I tilted my head toward the clouds, clutching my grip on reality as best I could.

What just happened? Wasn't it just a few minutes ago that I was suffocated by smoke and jolted by blaring fire alarms, whizzing bullets, and friends' shrieks of terror? Were there or were there not kids running and stumbling over each other while guys were shooting at them? Weren't teachers barking orders at us? Weren't SWAT team members peppering the school with return fire? Didn't my entire life pass before my eyes underneath that stupid table?

All I knew was that for now, all was still. And all was lost. I tried desperately to wrap my mind around the magnitude of what had just happened. *Protection was gone.* I was terrified to be alone and found myself spinning slowly in place, making circles on that field, my neck jerking ahead of my body with each revolution as I heard the distant *pop-pop-pop* of ongoing gunfire. *Laughter was gone.* My trembling lips, wet with tears and numb from shock, would surely never smile again. *Optimism was gone.* Who wanted to live in a world where this sort of catastrophe could happen? There was no point in having a positive attitude. Hope was just a pipe dream—a useless way to spend my energy.

*And my future was certainly gone.* How would I move on from this? I stood on that little patch of overgrown field, isolated and shivering in my damp, blood-splattered clothes, no possessions with me, no people by my side, and was kept company by a single thought: *My life will never, ever be the same.*

After April 20, so much was happening. At times I desperately needed attention and care, but other times I'd erupt in anger when someone asked a well-meaning question in an attempt to draw me out, soothe my pain, remedy my ache. I was dying a slow death, and the worst part was that I was watching it happen. I was terrified of being alone, but I hated other people's presence. I felt isolated and cold and empty, but I despised hugs and kisses and people squeezing my hand.

My friend Cassie had been killed in the library, and I couldn't make peace with that. Why hadn't I been the one to die instead of her? What was God thinking? How was I supposed to live with the fact that she was killed, while I survived—wasn't she the one who was living for God? I was angry and resentful and guilt-ridden, which drove me deeper into isolation, even from people I knew and loved.

If there were three words that plagued me after the massacre at Columbine, they were, *Why, God, why?* For everyone affected by that day, all that existed were questions. Eleven thousand pages of official police report later, there were still no real answers. At least not enough to outweigh the *whys* that plagued those of us still in the land of the living.

After the shootings, we had to attend classes at Chatfield for a few weeks. The teachers were so important—they supported us—we supported each other. There wasn't a lot of schoolwork going on, and there were things happening that brought fear on top of fear. But it helped for us all to be together. I will forever appreciate the teachers for showing up. We understood each other better than anyone else ever could. It felt so good to know that we had survived, and we knew no boundaries—we hugged everyone! We were just so grateful for being alive.

Life seemed to be a cruel joke at best. Stress took its toll. I was just getting by, just barely hanging on. Everything was a struggle. My mom and dad got a divorce.

I thought I would never heal emotionally, but of even greater concern was that I might never recover spiritually. Although I don't remember being angry at God, I was certainly angry at life for letting me down. And since I believed God was in control of the things that happened in life, I suppose my anger was inadvertently directed toward him. But I was too fearful to admit that. As the months went by, something in my soul kept pulling at me. I began to long for God again, even though I was angry and bewildered over all that he had allowed to happen.

After my graduation the next year, I knew I wanted something more than a summer of sitting by the pool or being tempted to party every night. I called a friend who had told me that if I were ever interested in making a difference in someone's life, he could make all the arrangements for me to work with a ministry called Samaritan's Purse. I had already been to war-torn Kosovo with the group, only eight months after the shootings. I believed that I could give back a portion of what the world had given Columbine by way of care, support, and gestures of compassion. This time, I went to Honduras, working with a group to rebuild following a massive hurricane there.

I had intended to become a teacher, but instead, I graduated from college and began another career track. I have gone to Breslan, Russia, to comfort the survivors of the terrorist attack on the school that left 330 people dead, 176 of them young children. I've traveled to Virginia Tech, Finland, Germany, Montreal, and Australia, reaching out to survivors of shootings and disasters there. I spent an entire summer in Africa, working with women and girls in a remote village in Mozambique. I authored a book, wrote a documentary for a film, founded a national speaking tour group, and have only recently stepped away, needing a little distance to allow space for other things in my life.

There is no doubt whatsoever that we are broken people living in a broken world. The proof may be seen in an angry teenager with a pistol or a religious radical with a bomb strapped to her chest or a drunk sitting behind the wheel of an out-of-control car. It may be an abusive father or an unfaithful spouse or an unreasonable boss or a friend who betrays you. It may look like the death of a dream, the death of a loved one, the death of a pet. But we all have experienced suffering and trials and disappointment that forever mark us. Maybe they differ in severity or impact, but they still exist—on an ongoing and relative basis—for everyone.

On April 20, I became a Christian under that library table, and that has sustained me. I have learned that there is an upside to suffering. I continue to seek hope and purpose in the aftermath of the tragedies I've experienced and witnessed. I have reached out to others, not because I enjoy public speaking, but because I realize that I have an opportunity to share messages of hope. People need to hear that good things can happen after a tragedy, and to see that it is possible to choose hope and discover purpose even after devastating loss.

\* \* \*

**WHAT'S NEXT?**

The assault on Columbine High School was by "two of its own." In the next chapter, you will read of a random act of violence in a school, the assault by a drifter on Platte Canyon High School. Marilyn Saltzman's account of this event focuses on the power of a small community and school attempting to heal from a shared sense of loss.

CHAPTER 8

# COMMUNITY AND SCHOOL HEALING TOGETHER

MARILYN SALTZMAN*

## THE INCIDENT

September 27, 2006, was a beautiful fall day, and the feeling was surreal—a déjà vu from the Columbine tragedy—as I headed down pine-lined Park County Road 43 in Bailey, Colorado, toward Deer Creek Elementary School. I had to dodge frantic parents crossing the two-lane road and circumvent the hastily parked cars as I searched for a safe place to leave my own vehicle. Dashing up to the school parking lot, I saw parents gathered in tight groups, waiting word of what was happening at the high school and anxiously anticipating the arrival of their children.

When I arrived, little was known except that a gunman was holed up in a room at Platte Canyon High School with female hostages. The rumors

*Author's Note: This chapter integrates information from personal experience, interviews, and material taken from the official report, *Lessons Learned: A Victim Assistance Perspective, 2006 Tragedy at Platte Canyon High School, Bailey, Colorado,* which I prepared for the Office of Victims Programs, Division of Criminal Justice Colorado Department of Public Safety (http://dcj.state.co.us/ovp/). Quotes from parents, students, teachers, and community members from that report are not attributed to individuals since the interviewees were guaranteed anonymity when the report was prepared. I would like to extend my deep gratitude to all who participated in this project. *Lessons Learned* was supported by Grant #2007-RF-GX-001 from the U.S. Department of Justice, Office for Victims of Crime, Antiterrorism and Emergency Assistance Program, January 2009.

were flying, and the air was laden with fear as parents watched for the school buses to arrive. Many of the parents had first assembled at the sheriff's substation in Bailey hoping to get the latest news. But information was slow in coming as the incident was unfolding, and finally parents were directed to the elementary school.

As the yellow school buses filed into the parking lot, parents searched the windows, frantically looking for their kids. Hugs and tears abounded as the teens disembarked, and families were reunited. Counselors from Platte Canyon and neighboring school districts were available in the cafeteria for families who wanted to talk about what had happened.

Slowly, the elementary school parking lot and adjacent road emptied of people and vehicles as families headed home. I stayed behind with school officials, piecing together the story and planning next steps. It was already dark when we joined the Park County Sheriff for a press conference.

"Unfortunately, tonight we are a community in mourning. Tonight, our hearts and prayers are with our students, staff, and their families, especially the family of the student that we lost. Our focus is on the health, safety, and well-being of our students, staff, and community members," said School Superintendent James Walpole at the press conference.

As the story unfolded, we learned that it was a senseless, random act of violence that shook a close-knit, trusting community and forever changed its face.

Platte Canyon High School is located right off U.S. Highway 285, outside of the town of Bailey, Colorado, about 45 miles southwest of Denver. The campus includes the school district's administrative offices, a public swimming pool, and a high school and middle school connected by a common corridor. Because of the swimming pool on site, visits by unfamiliar adults were a daily occurrence and no cause for alarm.

That Wednesday, a 53-year-old man parked his vehicle at the school in the morning, left briefly and then returned. It was later learned that he had been camping in the area and had no apparent ties to the school, the community, or the students. When he saw the school safety officer leave, at approximately 11:40 a.m., the intruder, wearing a backpack over a hoodie sweatshirt, entered the building, ascended the stairs, and entered classroom 206.

Claiming to have explosives in his backpack, the assailant ordered the teacher and all students except for seven girls to leave the room. When the teacher insisted that she would not leave without all of her students, he fired a warning shot from his handgun and forced her out of the room. As she left the room, she met Principal Bryan Krause preparing to enter. Just moments before, a student had told him there was an intruder in the building. A 911 call brought a crisis response team quickly to the school.

The high school and middle school went into immediate lockdown until classrooms could be evacuated in an orderly fashion. Following the school safety plan, teachers slid a green card under their classroom door if the classroom was secured, and the sheriff's deputies inserted their photo identification under the door, signaling the teacher to unlock the door for evacuation.

Students were first assembled in the gym and then followed law enforcement out of the building to waiting buses. Along with faculty, 770 students from the two schools were loaded on school buses and transported to Deer Creek Elementary School, ten minutes away, to be united with their families.

Over the next few hours, the gunman released five of the seven hostages. They conveyed that the assailant told them the crisis would be over by 4:00 p.m. Law enforcement assumed the intruder was going to detonate the explosives or take some other violent action. The decision was made to intervene, and at 3:35 p.m. the SWAT team entered the room after opening the locked door with explosives. One hostage ran from the class, was grasped by SWAT personnel, and taken to safety. The remaining hostage, 16-year-old Emily Keyes, was being restrained by the assailant, who shot her. She later died at an area hospital. The intruder was shot by officers and also sustained a self-inflicted gunshot wound to the head. He died at the scene.

It was a difficult and sleepless night for the students, staff, and parents of Platte Canyon High School. Their feelings of shock, sadness, and terror permeated the small mountain town. Law enforcement officers grieved together at the fire station after witnessing the loss of a student during the siege.

The next morning, school staff members met as a group at a local church and received details from law enforcement officers about what had happened. The school superintendent and the principal shared their grief and talked about next steps. Counselors and advocates from schools, the county, and community mental health agencies were on site to provide both individual and group counseling services as requested. The church also served as a community center on a drop-in basis, where students and their parents could gather to talk, support each other, and begin the slow process of healing. The Red Cross was on site to coordinate logistics.

As I drove to the church that morning, I was struck by the immediate and heartfelt community response. Pink ribbons and posters adorned the signposts along the highway. Local businesses displayed placards of support in their windows. Flags were at half-mast.

"What was helpful was the church being open all day and night to be able to gather with my peer group. They gave us all our meals because I couldn't even think about cooking. Having people around me made the biggest difference," a teacher said.

Within a couple of days of the tragedy, the Platte Canyon superintendent recruited a cadre of retired school administrators from the Denver metro area to call every parent in the school district to offer sympathy, assistance, and support. The administrators recorded a one-word reaction from each family so the school district knew how people were coping.

The high school and middle school were closed for five days to allow law enforcement to complete its investigation and to permit cleanup of the affected classrooms. The day before classes resumed, the school was opened for students and parents to retrieve belongings and gain a sense of comfort in re-entering the building.

Platte Canyon High School reopened on October 5, and about 45 mental health professionals from a number of state and local agencies were on call, with one counselor in every classroom and in the school offices.

The road to the school and the bridge connecting the school to the athletic fields were filled with pink ribbons and banners carrying such messages as "Be Strong" and "Random Acts of Kindness." A number of students prayed in front of the school before the day began, and students were given donated teddy bears and hand-knit scarves. Superintendent Walpole noted that of 460 high school students, only 10 were absent.

Once school resumed, parent and community volunteers manned the entryways to the schools, signing in visitors and providing badges. This effort not only provided more security for students, but also gave parents a positive, constructive way to help. "I liked having parents at our doors welcoming people rather than metal detectors and police keeping people out," one student said.

The high school started a program called Random Acts of Kindness, which encouraged students to be kind to others in Emily's memory. "It was a beam of light," a parent said. An annual event, near the anniversary of the shooting, still provides high school students with a choice of options for community service.

"It gets kids involved and acknowledges the good that could come out of it. Being kind to other people becomes a habit for teens," said Ellen Stoddard-Keyes, mother of the victim.

## A COMMUNITY COMES TOGETHER

A shooting in a small, rural mountain community makes every resident feel vulnerable because it is shocking in its unexpected, violent disruption of everyday life. "The women felt victimized; the men helpless," one community leader said. "We felt powerless if we couldn't even keep our kids safe," another said.

In a small community, there are multiple impacts because people work together, play together, and worship together. Lives intersect on a daily basis: The sheriff announces the high school football games; a sheriff's officer coaches the swim team; a school board member's spouse works at the school; a restaurant owner employed the victim.

Five years later, the Cutthroat Café in Bailey, where Emily worked, still displays a plaque near the entry with her photo and a poem dedicated to her. As her mother and I sat at a corner table in the café, a woman approached to give Keyes a hug.

The upside to a small, rural town is that intertwining, long-standing relationships build respect and trust. Because Bailey is a close-knit community, everyone—from the businesses to the churches to the schools—pulled together immediately. This cohesiveness was an incredible asset as the community moved from the tragedy to the healing process.

"The way that Platte Canyon handled this situation should be an example to the rest of the country on how a high school staff and community come together and support their students," a parent said.

The relationship between the Platte Canyon School District and the Park County Sheriff's Office was solid, so the two entities could work together seamlessly from the onset of the tragedy forward. The sheriff's office had run a practice drill in the high school just weeks before the shooting, and everyone knew exactly how to respond.

Principal Bryan Krause credits the drill with being critical to the smooth evacuation of the school because the faculty had so recently witnessed the exercise and knew exactly what to expect. He had almost canceled the event because they had done a similar practice the year before, and the beginning of school is always so busy. But he was glad he went forward with the exercise because the teachers were prepared for how the SWAT team would respond.

Park County Victim Services also had a strong working relationship with the school district, and a level of trust was already established, so victim-services personnel could provide immediate and ongoing support to students, parents, and school staff.

From offering meeting space for students and staff to holding workshops and providing gifts, the churches became integral in the healing process. Youth ministers at local churches offered a wide variety of activities for students and their families. A community vigil at a local church and an "Evening of Healing" at a nearby historic ranch are examples of events immediately following the tragedy that brought the community together.

Local professionals with a wide variety of skills offered their help. For example, students had the opportunity to take self-defense classes from a local expert. "He gave me back what that horrible man tried to take away from me," one hostage said.

The community gathered not only around the Keyes family but also created a protective circle around the six hostages who were in the classroom with Emily. All the students knew who the victims were but protected their identity from the media and the public.

Casey, Emily's twin brother who had been on a field trip the day of the shooting, "got a lot of strength from his classmates and teachers," said his mother. His class was less cliquish than so many high school classes, and he was surrounded by friends. His teachers even "spoiled" him, Keyes recalls, allowing him to set up his own one-cup coffee pot in the teacher's lounge that he could access at all times. Emily's teachers thoughtfully sent the family her English papers and artwork as a remembrance.

### IMMEDIATE MENTAL HEALTH RESOURCES

Within hours of the crisis, numerous local and state agencies responded to the scene to assist with mental health needs of the students, school staff, first responders, and the Bailey community. Under the auspices of the Park County Sheriff's Office, victim-services staff took a lead role in working with the victims of the tragedy, including the hostages and the first responders.

Written materials provided crisis information to parents, teachers, and students, including fliers explaining signs that indicate someone needs assistance and where to go for help. A resource packet, prepared by state and local mental health agencies, was distributed to all counselors who helped on the first day back at school and included talking points for teachers and handouts on trauma and post-traumatic stress.

"The services provided for the entire staff in the days directly following the incident worked the best for me. The presence of many available services was comforting. The group sessions where we vented as a staff were helpful. Services provided to the students involved were remarkable. I had a college-bound student change her major to psychology because she was so positively impacted by the services she received," a teacher said.

The school superintendent, victim-services coordinator, local government officials, law enforcement officers, and mental health representatives worked as a team to coordinate activities in the aftermath of the crisis. Two 24-hour hotlines, staffed by volunteers, were established immediately—one for people needing help, the other for those wanting to provide help. Fliers were distributed throughout the community at businesses, churches, and schools to advertise these phone numbers. A training session was organized for new volunteers, and a trained person answered the phone lines for several weeks. Later, an answering service took over with a "live" voice (rather than a recording) for at least six months. The hotline was "live" again for a couple of weeks around the first anniversary of the tragedy.

Other short-term services included a school district website that provided immediate, up-to-date information about events, activities, and where to go for help. This site was developed with assistance from Jefferson County Public Schools. A communications plan was developed through the services of a volunteer public relations consultant to inform parents about school district actions and decisions in a timely, accurate manner, and to involve the school community in events, activities, and safety issues. The sheriff's office provided an increased law enforcement presence in the parking lots and school buildings.

## OUTSIDE RESOURCES

The strength of a small rural community is counterbalanced by its lack of resources. In Bailey, both the sheriff's office and the school district run on lean budgets, and without the support of state and federal grants, they would have been hard-pressed to offer the ongoing services that were necessary to meet the needs of the community in the aftermath of the tragedy.

Mental health providers from neighboring jurisdictions and the State of Colorado were able to join forces with the Park County Victim Services staff to bolster manpower. Through interagency cooperation, a rapid, efficient, effective response was provided to meet the mental health needs of the community.

Principal Frank DeAngelis of Columbine High School called Principal Krause the day after the shooting to share condolences and offer support. He and his staff visited Platte Canyon soon after school resumed, bringing lunch and gifts for the Platte Canyon staff. "Talking to someone else who had gone through it was so valuable," a teacher said.

While help from neighboring law enforcement, mental health agencies, and school districts was welcomed by many, there was also some resentment regarding all the "outsiders in suits coming in to tell us what to do," a county employee said. Some of the first responders and local officials felt devalued by the "rescuers" from the big city. "People need to show respect for the locals," a sheriff's deputy said. "We got tired of all the people from outside trying to help," one school staff member said. "It was too much."

## FIRST THINGS FIRST

Principal Krause knew "right from the start" that his life, his school, and his community would be forever changed by the experience. As he describes it, he went from being a principal leading educational change to being a caregiver. As principal, he had set ambitious goals for improving student achievement, but after the tragedy, academics took a back seat to rebuilding the school community and shepherding it toward healing. "I needed to take care of the students, the teachers, and the community. There's no preparation for that. They don't teach you how to be a caregiver. It's a completely different job."

"It was essential to pull people together around a common sense of loss and address the emotions and feelings first. We couldn't assume a week later that we could go back to learning as usual," said Superintendent Walpole.

Teachers and law enforcement officers learned how important it was to strengthen their relationships with students and to share their pain with each other. "You can be human in front of the kids. Those teachers who did not attempt to be Superman or Wonder Woman positioned themselves to teach some real life lessons," a staff member said.

"My advice is not to underestimate the impact even on people who are not directly there. For example, there was a much bigger impact on our elementary students than we initially thought," Walpole said. Parents reported incidents of bed wetting and nightmares among the younger children in the community.

Children may exhibit "magical thinking," which allows them to feel more powerful, and therefore more responsible, than they really are, according to Dr. David Schonfeld, director of the Division of Developmental and Behavioral Pediatrics at Cincinnati Children's Hospital Medical Center and director for the National Center for School Crisis and Bereavement. He said children may wonder what they could have done to prevent the event and may feel guilty for no logical reason.

In leading the school after the tragedy, Krause made a conscious decision not to speak of the gunman again, only to look forward. To help reclaim a sense of safety he wanted to "create community again." Grant money helped pay for the immediate needs, including additional secretarial support that was needed to deal with the letters that flooded in. Some of the funds paid for substitute teachers, so teaching staff could take some time off if needed. With all of these strategies in place, Krause considers the strong school-parent relationship and the "amazing" close-knit nature of the community as the biggest factors in promoting recovery.

Teachers are on the frontline in a tragedy and need support to work with their students. Seeking out experts who could prepare teachers for what to expect was critical in the early days, Walpole explained. The teachers agreed.

"Knowing what to expect from my students was key. This prepared me with a plan to handle the manifestation immediately when it took place.

[A] student had a flashback of the SWAT team swarming into the room several months later and actually passed out. I was prepared to help take care of the fear of the actual experience," a teacher said.

The superintendent and his staff learned to lean on community groups in ways that school districts don't often consider. "We had to broaden our discussion from how the school does things to interact with the religious community, organizations, and parents, and ask for their help," Walpole said.

Walpole also noted that in addition to dealing with the emotional and academic impacts of the tragedy, the school district had to immediately deal with the business issues that such an incident creates. Lawyers and insurance agents were involved from the onset. Construction workers were needed to repair the damage caused by the gunman and SWAT activity so that school could resume.

## IMPACT OF MEDIA ATTENTION

Local media, learning about the incident at Platte Canyon on their police radios, appeared at the scene almost immediately. They were joined the next day by national media, including CNN. A media camp was set up on the football field across the highway from the school.

As in all school crisis situations, officials had to prepare for a swarm of immediate and ongoing media attention. "There were TV cameras parked in front of our house for weeks—they wanted to talk to my children," said Principal Krause.

The primary goals in dealing with the media were to provide accurate, timely information to students, staff, parents, and the wider community; to help the community heal; and to protect students so they could return to school with a sense of safety and normalcy.

In the initial hours after the crisis, the media was consumed with finding out what happened by interviewing officials as well as anyone else willing to talk—students, staff, and parents. As the hours and days passed, the focus turned to why the incident happened and more in-depth stories about the perpetrator and the victims.

While some Platte Canyon students were eager for the media spotlight, others just wanted to grieve in privacy. One young man concocted a story of his own heroism (perhaps wishing that he had been able to act that way) that garnered national media coverage. The young man later admitted his story was a fabrication, and the press had to retract it.

"We worked well with the media because we chose our interactions," Keyes said. "We've put a positive light on everything we've done with the media. We specifically have requested that they didn't mention the perpetrator's name or use his photo." The goal was to remember and honor Emily rather than focus on her assailant, and families in subsequent shootings have followed their lead.

To protect grieving staff, students, and community, media were not permitted inside the church where mental health services were being provided. Reporters were required to stay at the media camp and not venture into the school parking lot in the days immediately following the tragedy.

A flier, *Your Rights with the Press,* was distributed to students, staff, and parents. The same flier, prepared by the Victim Assistance Unit of the Denver Police Department, had been distributed at Columbine. It outlines victims' rights, including, most importantly, the right to say "no" to an interview, the right to be treated with dignity and respect, and the right to grieve in privacy.

It was important to hold ongoing press conferences and issue press releases from both the sheriff's office and the school district to meet the media's need for updates while protecting the integrity of the investigation and the privacy of the victims. Whenever new information became available—either about the crime or about school activities—the media received it.

On October 3, two days prior to reopening the school for classes, the school district offered an escorted media tour led by Superintendent Walpole. Cameras were permitted in the building for that one event only. Reporters were not permitted to enter Classroom 206 where the incident occurred. By providing the tour, the district granted media requests for access to the school building when it would not interfere with student learning.

The school district also prepared for media attention at benchmark events, such as graduation and the one-year anniversary of the shooting. Media representatives were permitted to attend the graduation in June 2007 with strict guidelines that included designated parking and seating areas. Media were asked to wait until the conclusion of the formal graduation program before approaching parents or students for interviews.

As the one-year anniversary of the tragedy approached, the school district sent a news release announcing that school commemorative events would be closed to the media, but there was a public event that evening. The news release noted,

> As you probably know, mental health experts say that when victims view images from a tragedy they experienced, it has the potential of re-traumatizing them. Therefore, we are asking for your help in avoiding replaying the emotional footage from September 27, 2006. We would truly appreciate it if instead, you mark the event with positive images, such as our graduation or other healing events. We also respectfully request that helicopters not fly over our schools at this time.

The purpose of this message was to reinforce Platte Canyon's long-term media strategy—to promote healing, to return to some semblance of normalcy, and to prevent re-traumatization to the greatest extent possible. This strategy was developed as a result of the experience in schools where other shootings had occurred, including Columbine, where the media attention was overwhelming and sometimes intrusive to the students, parents, staff, and the community.

### LONG-TERM SUPPORT

While there was an abundance of support in the immediate aftermath of the tragedy, mental health services were needed for months—and even years. When a loss is so public, the grieving process is delayed because victims are dealing with media and public attention rather than attending to their own feelings. There is no deadline for when grief begins or ends, especially when related to post-traumatic stress. The challenge is that eventually funding runs out, and it becomes harder to maintain the professional support systems that were initially in place.

"You don't get over it. Life-changing events change lives. They become part of your life," Schonfeld told Platte Canyon parents at a workshop in early December 2006, about two months after the incident. He noted that there is no timeline for how long recovery might take, as it varies by individual.

"My advice is to secure long-term funding as quickly as possible. Even for two to four years," said Keyes. Her own post-traumatic stress symptoms took the form of physical illness—sciatica and digestive troubles—and manifested from months to years after the tragedy. She noted that five years later, she knows of some of Emily's classmates who are just beginning to deal with their grief as adults.

Parents, staff members, and students all expressed the need for long-term support. "The process of recovering from an event of this type will be ongoing. You may not need help right away, but may three months, six months, or a year later," a staff member said.

"Remember that it [reaction to tragedy] first can appear months after the event and that it may be difficult to attach those fears to the event. Watch for emotional distress for a long time afterward," a parent advised.

"I found that after six months it seemed too late to actually talk to somebody, and that's when I really needed to talk to somebody," a student said. Another noted, "I was in denial for a really long time after it happened. I thought if I didn't think about it, I wouldn't be hurt, but as soon as school was out for the summer, I got really depressed, and everyone else was already over the worst of their grief, so I felt like no one understood me."

"Although my counseling services were terminated, it would still be helpful to feel like we had a time and place to talk about the lingering effects this had on us," one teacher said.

Long-term support is especially needed around the anniversary of a tragedy. Park County Victim Services worked with the school district and community groups to plan events to commemorate the first anniversary of the shooting. Elementary and middle school students did projects with the theme of "acts of kindness" at their school. Participation was voluntary, but attendance was excellent. The high school students participated in a wide range of community service activities, culminating in a picnic. In the evening, there were a community bonfire, barbecue, candle-lighting ceremony, and wall of remembrance where people could write their hopes and thoughts.

One long-term support system put in place was the Platte Canyon Drop-In Center, which opened in February 2007, to provide students with a safe and nurturing environment. Located in downtown Bailey, it served middle and high school students with activities and a safe place to congregate after school. The school district provided free transportation from the high school. Unfortunately, after a few years, the center closed because of a lack of funding. A Boys and Girls Club has been formed in its stead, but it too struggles with a constant dearth of money.

### SAFETY AND SECURITY

"As a school administrator, you don't get over it. I worry about security every day, way beyond how I ever did before," Walpole said.

He noted that immediately after the incident, it was important to put measures in place that not only increased safety, but also made people feel safer. Examples included limiting access to the building by locking doors, increasing law enforcement presence, and having parents in the school lobby to greet students, staff, and visitors.

In mid-November 2006, approximately 70 Platte Canyon staff, parents, students, and community members attended a safety forum to learn more about what the district was doing to enhance safety and to provide feedback on potential new safety measures. An online survey asked community members to rate safety measures they wanted to see the district enforce.

In January 2007, the Board of Education convened a safety committee that met every Saturday morning over a nine-week period to study school safety across the nation, identify concerns, and make recommendations for their schools. The group developed twelve recommendations, including more school resource officers, video surveillance cameras, limited access, and ID badges. However, there was no consensus among the committee members about priorities. While the community had come together to heal from the tragedy, it was more factionalized over the issue of future safety measures.

"There were pretty heated discussions in the community. We had one group that wanted high security and another group that didn't want anything but student awareness training. The Board decided what actions to take, and the issue settled down," Walpole said. The result was a middle ground, where some new measures, such as limited access, remained in place.

### KEYES FAMILY AND THE I LOVE U GUYS FOUNDATION

Whether interviewing educators, community members, or parents, I heard the same message repeatedly. Because of the positive and gracious response of John-Michael and Ellen Stoddard-Keyes, the community healed more quickly.

"The Keyes set an example of kindness, not hate," a parent said. "They facilitated our healing," a community member said.

"Time does heal, and trying to better your life and that of others *because of* what happened—a lesson Emily's parents taught us—is a tremendous way to truly heal," a teacher said.

"We set a tone and the community followed it. It is a tone of love and not one of hate," Ellen Keyes said.

The Keyes family immediately turned the tragedy into a force for community kindness, from the donation of Emily's organs to the founding of the *I Love U Guys Foundation* to provide support for positive community activities. The name of the foundation comes from Emily's last text message to her family as she was being held hostage in her honors English classroom.

The statement on the *I Love U Guys Foundation* website captures the philosophy that the Keyes family has adhered to from the very first day:

> For some it may be difficult to accept. The right actions at the right time by the right people may not have the right outcome. When what we hoped for didn't happen, some may want to find fault. But there is no fault to be found in the command decisions made, given the information and behavior presented. There is no fault to be found in the courage and speed of their response. From our family there is only respect.

Just 10 days after the shooting, the foundation sponsored the Columbine to Canyon Ride—*Emily's Parade*. The impromptu motorcycle event drew more than 5,000 riders and raised $64,475 from donations. The funds were donated to the six surviving hostages and their families.

The motorcycle ride and 5K run/walk, held annually around the anniversary date of the tragedy, are powerful because they bring together families, law enforcement, and two communities that suffered unspeakable loss. Starting at Columbine High School, located in a Denver suburb, the motorcycle ride winds up through the foothills to the rural Platte Canyon High School. Community members line the highway along the route to show their support.

The fundraisers help finance *I Love U Guys Foundation's* programs, which include annual scholarships to Platte Canyon graduates and a Kindness Award to a Platte Canyon teacher.

Since its inception, the foundation has evolved to developing safety programs for schools around the nation. John-Michael Keyes has given more than a hundred presentations nationwide, and as a result of his travels, discovered there was a lack of common language for crisis response. Working with experts he had met across the nation, Keyes wrote the Standard Response Protocol, a response to any crisis scenario that includes four specific actions. Jefferson County Schools, the largest school district in Colorado and home to Columbine High School, has adopted the protocol as have school districts across the country. The protocol is used as an addendum to school safety plans. The four actions are as follows:

- Lockout—safeguard students and staff in a school building by securing the perimeter.
- Lockdown—secure individual rooms and keep students quietly in place.
- Evacuate—move students and staff in or out of a school building, with specific instructions about the new location.
- Shelter—take a protective position, a protocol followed by a type and method of protection, such as move to a hallway, or drop, cover, and hold in case of tornado.

"Emily gave us a voice so we can talk about this. If we stay silent about the reality of students being in danger, we wouldn't do any service to anybody," Ellen Keyes said. She admits, however, that "telling our story is like picking a scab and opening it over and over. But using the emotional part of our story gets people's attention, and we end on a positive note, giving them tools they can use."

The foundation has also taken on a statewide advocacy and awareness role. The Keyes family worked with the Colorado Legislature to bolster interest in organ donations through the Emily Maureen Ellen Keyes Organ and Tissue Donation Awareness Fund. The foundation also sponsors awareness campaigns related to teen sexting and the "Regretting is Harder than

Telling" campaign and curriculum to encourage youth to report potential incidents of school violence.

## ADDITIONAL RESOURCES NEEDED

When a crisis occurs, school districts, whether large or small, will have to depend on the assistance of outside resources, both human and financial, to manage the tragedy. As part of crisis planning, school districts, law enforcement, and victim-services agencies should conduct an assessment of what resources are available in the community and what might be needed in the advent of a tragedy, with special emphasis on serving victims. Helpful personnel include

- substitute teachers to replace staff who are unable to return temporarily, or to give teachers a break when needed;
- additional secretarial help to handle phone calls, record gifts and donations, and keep track of resources and activities;
- mental health counselors and nurses at school and in the community;
- law enforcement and security personnel on site to handle visitors and the media, and to meet needs for increased safety;
- trauma experts who can make presentations to students, staff, and community on dealing with trauma and what to expect;
- public relations professionals who can develop communication tools for parents, staff, and the community as well as handle the onslaught of media attention; and
- grant experts to attain and manage government and private funding sources.

Funding may also be needed for such unexpected costs as building repair and reconstruction, safety and security hardware such as new locks and video surveillance cameras, and financial assistance to victims. Potential sources of funding include federal, state, and municipal grants; school district insurance; school contingency funds; school district capital funds; foundation grants; and private donations.

National organizations such as the National Education Association, National Organization for Victim Assistance, National Office of Victims of Crime, U.S. Department of Education, American Counseling Association, National Association of School Psychologists, National Center for School Crisis and Bereavement, and National Center for PTSD, have materials that are extremely useful to teachers in dealing with student and family issues after a crisis.

Often after a tragedy, people near and far wish to contribute to the school or to the victims. I remember sitting in the Platte Canyon administrative office late on a Friday afternoon when a young man who was driving by on Highway 285 stopped in and handed the secretary a $20 bill. "I just wanted to do something," he said.

To handle the outpouring of support, it is critical that good tracking procedures are established immediately to ensure the funds are used appropriately, and there is accountability for how monetary donations are spent.

### SOME LESSONS LEARNED FROM PLATTE CANYON

- In a small community, a school tragedy affects everyone, and adults and students can draw strength from each other through special events and activities.
- Schools should foster strong relationships with local governmental agencies such as police, fire, and mental health, and develop crisis plans together.
- There is no deadline for healing. Every individual heals at a different rate, and schools should plan for long-term mental health support.
- Small communities may not have the resources to handle a school crisis without outside help. The crisis plan should identify neighboring entities that can provide support.
- Teachers are on the frontline and need tools to help their students heal and return to a sense of normalcy, so they can return to learning.
- The impact of the media on student healing must be considered. Work closely with the media to meet their needs and help them understand yours.

### CONCLUSION

The random, senseless attack on Platte Canyon High School by a deranged intruder bent on self-destruction violated not only the safe, sheltered environment of the staff and students, but also that of their parents, friends, and neighbors. The effect was traumatic and confounding for the residents of this bucolic Colorado community, previously best known for its good fishing, camping, and breathtaking scenery.

To heal from a violent and staggering intrusion, it took the entire community's supporting the school and each other, as well as financial and human resources from neighboring communities, the State of Colorado, and the federal government. Immediate and long-term efforts, including mental health and safety measures, were critical to bring a new sense of security and normalcy to the students and staff at Platte Canyon High School. Teachers and administrators had to obtain new tools and learn new skills to deal with the effects of trauma on their students as they re-engaged them in academic pursuits.

While the long-term impacts cannot be overstated, by working together, a community can emerge stronger after a tragedy.

### CONTRIBUTOR NOTES

**Marilyn Saltzman** worked in the Communications Services Department of Jefferson County Public Schools, Colorado, for 20 years and was communications manager in 1999 during the time of the Columbine tragedy. She was on scene at Columbine within an hour of the tragedy's onset, handling the media and other communications needs. Because of her experience at Columbine and because she lives in the foothills of Colorado near Platte Canyon High School, she was called on the day of the shooting there to help with communications efforts. She continued consulting with the Platte Canyon School District for the next year and was later asked to prepare the *Lessons Learned* report.

### WHAT'S NEXT?

In Chapter 9, Nina Lyytinen and Kirsti Palonen share the model for mental health services that were offered in school and throughout the community after a shooting in Finland. Lessons learned from the tragedy were employed after another Finnish school shooting one year later. Through individual and group assistance, the aftercare services fostered resilience and helped affected individuals to psychologically reclaim their school and their life.

CHAPTER 9

# AFTERCARE: SUPPORT FOR SCHOOL PERSONNEL FOLLOWING A SHOOTING IN FINLAND

## NINA LYYTINEN & KIRSTI PALONEN[*]

### BACKGROUND

On November 7, 2007, a shooting at the Jokela School Center shocked the entire Finnish nation. Jokela, located in the northern center of Tuusula municipality, is a community of 6,000 residents. The School Center houses upper school (grades 7–9) and high school. At the time of the shooting, 489 pupils and almost 50 teachers and staff were in the building.

Just before lunch time, an 18-year-old male student opened fire, killing six students, the school principal, the school nurse, and himself. The gunman moved around the school, shooting and shouting. He also attempted to set fire to the building. In addition to those killed, one person was injured by the shooting, and eleven were injured by shattering glass as they attempted to escape. Most of those in the building were exposed to this psychologically traumatizing event.

---

*Author's Note: This chapter is dedicated to the Jokela School Center personnel, in deep appreciation of their professional knowledge and skills. School personnel showed a great commitment to their work and to taking care of the well-being of students in a situation for which there was no previous model in Finland. Already highly educated, they mastered what amounts to "a degree in trauma survival" following the disaster.

Prior to this incident, the only other school-related shooting in Finnish history had occurred in 1989, when a 14-year-old student killed two fellow students at the Raumanmeri School in Rauma. However, less than a year after the Jokela shooting, tragedy struck again. On September 23, 2008, a male student opened fire at the Seinäjoki University of Applied Sciences in Kauhajoki, about 300 km (186 miles ) from Jokela. The shooter killed ten people at the university before killing himself.

Following the Jokela and the Kauhajoki shootings, hundreds of threats were made to schools throughout Finland. As a result, school communities experienced increased feelings of vulnerability and risk.

In Finland, after a disaster or catastrophic event, psycho-social support is typically provided as a means to foster the resilience of communities and individuals (see Saari, 2005). It aims at easing the resumption of normal life, assisting affected people in their convalescence, and preventing pathological consequences of potentially traumatic situations. To date, there has been little documentation of the effectiveness of psycho-social support provided to school personnel after a shooting. This chapter, based on a follow-up study of the Jokela shooting (Lyytinen, 2010), addresses the psychological effects of the tragedy on educators and school staff and considers changes in their symptoms in the year following the event. Additionally, it offers suggestions for responding to such a tragedy and considers the perceptions of those receiving such services.

Although what was learned at Jokela informed the response to the Kauhajoki shooting, it should be noted that the two situations were actually quite different. At the university in Kauhajoki, students, families, and faculty were from many different locales, but in Jokela, most students, families, and school personnel were from the local area. As a result, the entire Tuusula region (where Jokela is located) was greatly impacted. The event affected health care systems, the school system, families, and individuals throughout the area, because "there is always someone you know." Recommendations for assisting school personnel in the aftermath, provided at the end of this chapter, incorporate what was learned from both the Jokela and Kauhajoki tragedies.

## WELFARE OF SCHOOL PERSONNEL AFTER A SCHOOL SHOOTING

Schools play an important role in the community, for they are responsible for educating and nurturing children into adulthood. School personnel who are exposed to lethal school violence are at risk of developing post-traumatic stress disorder (PTSD), and after such an event, the teachers' role becomes even more important than before. Students who return to the school need adults to reconstruct a safe environment for growth and learning. Students may lean on their teachers after a shooting because they shared the experience. However, at the same time that the teachers are supporting their students, they are laboring to reconstruct their own sense of basic safety and professional identity.

In spite of the increased attention on them and on their performance, school personnel often feel neglected following a school shooting (Newman, Fox, Harding, Mehta, & Roth, 2004). After a traumatic event, teachers are called on to evaluate changes in students' behavior and to plan instruction to meet changed needs. Since teachers know the students' learning and concentration abilities before the incident, they are best able to provide important information about changes in learning, attention, and other behaviors. Teachers are asked to support mental health interventions for their students and to reassure parents after the incident. However, these are enormous expectations of them since they are victims themselves and need additional support in this demanding role.

Following a traumatic event, educators are responsible for teaching traumatized children, which means that their duties are now comparable to teaching special needs students. Modifying curriculum and instruction to meet special learning needs is not a simple process even in routine circumstances. It poses an even greater challenge for teachers who themselves have been traumatized, and who may not know how to make necessary curricular adjustments. Teachers who may not have been trained to provide mental health support for their students need assistance in meeting these new challenges, understanding normal crisis responses, identifying behaviors that are cause for concern, and gaining access to professionals for guidance.

Individuals who have experienced traumatic situations often resist treatment and hesitate to consider themselves as patients. Yet, untreated traumatization can create a high risk for PTSD. Therefore it is important to evaluate school personnel after incidents of school-related violence, not only to encourage them to seek counseling but also to help them identify personal strengths, resources, and strategies that have helped them cope with challenges in the past.

## PSYCHO-SOCIAL SERVICES FOR SCHOOL PERSONNEL

Immediately after the Jokela shooting, the municipality of Tuusula formed an administrative group charged with planning the aftercare response to the event. Aftercare services include all support that is provided for people who are exposed to a traumatic situation, for the purpose of helping them to restore their quality of life and functioning ability. The aftercare management group included personnel from all sectors affected by the shooting (educational services, youth services, health care, occupational health care, social services, family guidance center, and financial management). Their role was to assess needs, plan, monitor, and evaluate services. Within a week, the aftercare group was reinforced with experts in crisis and catastrophe work who functioned as the aftercare coordinators. Their experience from previous catastrophes contributed to the planning process, and it was planned that aftercare would extend through 2012. The Finnish government provided financial resources to the municipality of Tuusula to implement the aftercare plan.

An aftercare coordination group included program coordinators, two psychologists/psychotherapists, and two crisis workers to function in cooperation with the municipality's aftercare management team. This group provided psycho-social services and consultations for those who experienced the shooting. They assisted school personnel in making aftercare arrangements, such as finding a psychotherapist and applying for compensation from insurance companies, social insurance institutions, and the State Treasury. Aftercare coordinators helped plan intervention services and also held educational sessions at the school to teach the participants about the effects of the traumatic experience and to provide strategies for discussing crisis and trauma reactions with parents and others in the community. The coordinators also worked with the media to educate the community about normal crisis responses, availability of services, and related topics.

Crisis work at the Jokela School Center started as soon as school personnel and students were evacuated from the building. Psychologists and crisis workers provided psychological first aid at a crisis center set up in a nearby church. The day after the shooting, the Finnish Red Cross (FRC) Psychologists' Preparedness Group provided psychological first aid for school personnel through group intervention, a process called *defusing*. Thirty teachers participated in the session. By the following day, other group sessions were held to allow participants to meet with colleagues to talk about their experience and to begin articulating what had happened. Group sessions served as a place to build a story of the event based on facts, not rumors.

Group processing of a traumatic experience is considered highly therapeutic for people exposed to a collective trauma. The protocol for the group sessions was based on the FRC Psychologists' Preparedness Group's agenda for work with trauma-exposed groups. A single group meeting is not always sufficient, so multiple sessions are often needed. The process of forming groups for the psychological intervention was guided by the central principle that individuals in a group must have experienced a common stressor. In structured sessions, participants review in detail the facts, their thoughts, feelings, and reactions to the event. Participants are given information about normal reactions to a traumatic event and helped to increase their knowledge of coping methods.

All school personnel at Jokela School Center were offered group support within the first four days after the shooting. The 42 partcipants who accepted assistance were arranged into four groups, based upon their level of exposure:

1. Those who had been in mortal danger, had seen others being shot, or had seen dead bodies.
2. Those who had been in the building when the events occurred.
3 & 4. Two groups of personnel who were not in the building when the events occurred.

Over the course of two months, between two and seven sessions were held for the groups. The least exposed groups met twice, and the group with the greatest level of exposure met seven times. After the group sessions, participants in need of additional care were referred to individual counseling. Additionally, counselors provided support via telephone when needed.

Of the participants who responded to the follow-up study, 87 percent perceived that the psychological group intervention was helpful. They thought that "it was crucial for their survival," stating that they felt cared about and that it gave them hope that they "would survive." Participants also said that it was important to have intensive support in the beginning and that the support continued across several group meetings. Only two participants who had extreme exposure felt that hearing other people's experiences was not helpful. They preferred family support and individual counseling. Two participants who were present at the school but not directly involved said that they did not benefit from the sessions. School personnel who were not at the school on the day of the shooting found the psychological group intervention helpful. The clinical experience that most participants found supportive was the psychological group intervention.

### RETURNING TO SCHOOL

School started again on November 12 (five days after the shooting) at a temporary facility, and for three days extra support services were readily available. Psychologists from the FRC group provided support via individual and group counseling for students and school personnel.

The day before teaching resumed at the Jokela school facility, school personnel, accompanied by FRC psychologists, visited the school, which was renovated after the shooting. School personnel walked around and checked every corner of the building to feel that it was safe for them to work there. The goal was to psychologically take possession of the building.

On November 15th, one week after the shooting, school personnel and students were able to return to their own building. Their school had lost two central members of the staff: the principal, who was the leader of the school, and the school nurse, who was a central part of the student welfare group. Losing a principal means that teachers have lost their administrator and their colleague. Even in a nontraumatic situation, such a loss brings feelings of grief and insecurity within a school community. Especially when a loss results from a tragedy, the school needs a strong leader, capable of providing support, direction, and guidance for working toward a daily routine appropriate to the acute situation. It is also important that the leader knows the school system, the municipality, and has strong networks in the affected area. It was beneficial for the continuity of school work at Jokela School Center that a principal of another school, who was a central member of the Tuusula school system and municipality, was chosen as a substitute principal to lead the school through this difficult time.

The substitute principal, vice principal, and two crisis psychologists worked together to plan the return to school. On the first day of teaching, 24 psychologists and crisis workers were on hand to offer support for staff and students. Each class had a support person available for the teacher and the students. Also, 10 substitute teachers were hired to ensure that classes would not be interrupted in the event that any of the regular teachers could not carry on their work.

Intensive psychological support was available at the school for a week after the return. The municipality of Tuusula also hired two experienced crisis psychologists to work at the school: One worked primarily with the students through the end of 2007; the other worked primarily with school personnel through the end of 2008. Low threshold therapeutic services were provided for students and their families until 2009 at a location near the school. In addition, health care and psychological services were available to the personnel at the school and at facilities nearby. Since the psychologist working with school personnel was based at the school, she was easily accessible and became like a member of the school faculty. She was also available to assist in parent conferences when needed.

Some school personnel felt that they needed more support than was available on the day classes resumed at the School Center. Teachers especially wanted guidance on how to face the students and how to take the traumatic experience into consideration when returning to academics. Even though there was one psychological support person for each class, some teachers found this insufficient. The situation revealed that

- a school needs more support personnel when resuming classes after a traumatic event;
- school personnel need more information about likely student reactions;
- educators need strategies for handling difficult class situations when they are also under extreme stress; and
- those affected by trauma need to accept that in order for the interventions to be helpful re-experiencing some of the traumatic reactions in a therapeutic context is unavoidable.

Administrators and district managers may have difficulty understanding why so much support is needed, especially in Finland since teachers there are required to have master's level degrees and are expected to be able to deal with whatever comes up in the course of teaching. However, after a shooting, the entire school community has lost its sense of basic safety and normalcy. This situation leaves the teachers, who know how to teach in a normal situation, alone, with the fear of possibly breaking down in front of students or colleagues. It is a great challenge for school personnel to handle their responsibilities when they have faced such extreme and unexpected stress in their work place. Teachers in Jokela had no peers at other schools in Finland to ask how to manage the challenges they faced.

## ADDITIONAL SUPPORT

Even though it was not statutorily required at that time, Jokela school personnel were granted part-time sick leave for a period of about two months, a resource that was viewed as "beneficial for survival" by the staff. Many, however, wished for a longer period of leave since their return to work was very difficult. Currently, part-time sick leave in Finland means that the workload (number of hours/week) is reduced to 40–60 percent of the full-time work schedule (Kela, 2011). After a traumatic event, a return to daily routine can support recovery; however, a great deal of energy goes into processing reactions to the trauma. Easing into the workload allowed personnel to return to work in smaller steps. A school thus needs to employ sufficient temporary staff to cover the work of personnel who are taking leave time. Another option for easing the return to school is to shorten the work/school week, a model that was used after the shooting in Kauhajoki.

After the shooting, occupational health care services were strengthened. Tuusula hired a doctor (part-time) to provide extra assistance to the Jokela area and an occupational nurse to visit the school regularly to conduct health checkups and provide other services. In addition, an occupational psychologist was available to school personnel. Health checkups were perceived as particularly helpful, as some of the personnel, for example, did not realize how high their blood pressure was until their checkup. School personnel who made use of these services perceived them as beneficial, but felt that still more support was needed.

School personnel were also offered crisis counseling and long-term psychotherapy. Crisis counseling typically involves one or more sessions to address immediate reactions and education about normal responses to trauma. People who are faced with a crisis situation often find it helpful to learn about normal stress reactions. Crisis counseling is generally offered by mental health professionals, such as psychologists or psychiatric nurses who have been trained in crisis intervention.

After receiving crisis counseling, some staff members pursued long-term psychotherapy, which usually lasts from one to three years. It is not uncommon for school personnel to need therapy several years after a school shooting. Therefore, survivors need to understand the long-term effects of exposure to traumatic situations and realize that some may not pursue long-term therapy at all or might not begin therapy until years after the incident.

In Finland, eligibility for financially supported psychotherapy requires a three-month psychiatric evaluation and a trial treatment. Kela, the social insurance institution of Finland, provides access to psychotherapy for one year at a time, for a maximum of three years. This insurance financially supports a certain amount of the cost of psychotherapy, and the individual pays for any charges over that amount (Kela, 2011). After the Jokela shooting, school personnel were offered crisis counseling with the costs covered either by occupational accident insurance or by the municipality of Tuusula. In most cases for the long-term treatment, the municipality covered the cost of the psychiatric evaluations and the amount that was not covered by Kela or insurance for the psychotherapy. They also paid for travel expenses to/from therapy. Personnel were encouraged to apply for additional compensation from the State Treasury for other losses (e.g., for broken personal items, or psychological and physical pain resulting from their experience).

In some cases, procedures involved in applying for and gaining access to affordable individual trauma counseling are difficult to negotiate. Steps that may, under normal conditions, be easy to complete can be difficult for traumatized individuals, and even filling out forms for compensation or services can be overwhelming. Official procedures for requesting help should be simplified as much as possible, and clerical help may be needed to assist victims with the requisite forms and to distribute information about compensation and access to services.

Most of the school personnel who participated in the follow-up study found the crisis support and psychotherapy beneficial, noting that it had helped them understand their own actions and reactions after the traumatic event. They mentioned that "the professional therapist gave skills to survive." Some stated, "Without it I might not have been able to keep the ability to work," and "I would not have survived without [it]." Psychotherapy also helped them deal with their reactions to the building where the shooting occurred and other stressors arising from the crisis, including difficulties in their personal life. For some, a crisis situation can interfere with other parts of life and strain close relationships.

Study participants considered Eye Movement Desensitization and Reprocessing (EMDR) (Shapiro, 1989, 1995) helpful in reducing trauma response symptoms and reducing flashbacks and intrusive memories. Participants also appreciated the opportunity for discussions with others. For those who started treatment immediately after the event, PTSD symptoms decreased.

## GETTING BACK TO TEACHING

Following a school shooting, teachers are acutely aware of themes of violence and death in the subjects they teach. For example, in a subject like history, literature, or religion, the theme of death and dying is not uncommon. Teachers may begin to avoid lessons on these topics in an effort to protect the students and themselves from the pain their loss has caused. It is important to provide guidance and support in how to deal with these curricular issues.

Work counseling groups were formed to help school personnel fulfill their job responsibilities. Counselors organized interested school personnel into five groups, but the plan did not produce the kind of benefits that were hoped for. Some personnel were disappointed in the work counseling because of the group formation and because the sessions began by going back to the day of the shooting. The group formation was experienced as a problem because each group included a variety of staff who worked in different areas of the school (e.g., teachers, cleaners, cooks) and who therefore needed different types of work guidance. Returning to the events of the day of the shooting was problematic because personnel had already participated in psychological group interventions that served as support in dealing with the shooting and their reactions to it. Three of the work counseling sessions were discontinued, and two groups continued with the original plan. The ineffectiveness of this approach to supporting school personnel in the aftermath emphasizes the importance of careful formation of groups. It is also important to plan sessions so that they meet the overarching goal of assisting people as they return to their daily work. While professionals working with trauma survivors need to hear their experience, it is painful for the victims to have to repeat it. Counselors need not only to ask victims to discuss their feelings about a traumatic event but also to discuss the reactions that arise as they repeat their stories.

Psychological trauma evokes intense sensory impressions, and bodily tensions are common. Treatment methods that take into consideration the physical aspects of the traumatic experience, such as relaxation techniques, can be effective in balancing the physical and psychological stress reactions as well as offering avenues for self-care. A physiotherapist provided individual and group treatment sessions for Jokela school personnel experiencing physical tension following the shooting. Also a music therapist conducted a 10-week relaxation class for school personnel. Coordination of the treatments of the physiotherapist, music therapist, and psychologists was beneficial. Participants felt physiotherapy helped them reduce their tension, understand their physical reactions, and realize the essential connection of body and mind.

The shooting affected those exposed at the Jokela School Center, those in other schools in the area, and local residents as well. As a result, the aftercare group involved members of the municipality's administrative level in planning services as well as in authorizing financial resources. Information about psychological effects of the trauma was shared through the local media and the Internet to the whole community. Articles in local newspapers and on the municipality's website explained normal reactions to crisis situations and where to go for aftercare services. The educational services of the municipality of Tuusula began a comprehensive development project to help alleviate any future impacts of the Jokela shooting. Through this project more school personnel were hired, training was provided to personnel, and the cooperation between school and home was strengthened.

### SOCIAL SUPPORT

Social support is often defined as the availability of people on whom one can rely, people who let one feel cared for and valued. Success of this type of support depends on two basic elements: (1) the perception that there is enough support available, and (2) the degree of satisfaction of the perceived support available (Sarason, Levine, Basham, & Sarason, 1983).

In the aftermath of a school shooting, some survivors feel that social support declines too soon, since family members and others in the community do not understand the long-term effects of such an experience. Survivors feel they are expected to get over their trauma quickly and if they are not able to do so, that something must be wrong with them. Some feel that the shooting is a forbidden topic of discussion. It is indeed difficult for those who have not experienced such an event to talk about it, and they often fear that they do not know how to provide appropriate support. Some survivors simply stop talking about the events when they notice that others try to avoid the subject.

In the response to the Jokela shootings, teachers and staff received social support from family, friends, co-workers, and professional health care providers. School personnel felt that family and friends helped them survive. The opportunity to talk with them and to know that they would listen and not exaggerate the event was beneficial. They felt that family members helped them return to normal daily routines, a factor that supported their recovery so that the experience did not keep them from doing what they were used to doing. Children and spouses were seen as vital supports. Home was a place where they could talk, share, and play. Family members "knew you before the trauma" and therefore it was easier for them to understand the challenges that were being faced.

School personnel also felt that support from their colleagues was important. Having faced the same traumatic event, they were struggling with the same challenges at work. Colleagues provided understanding that others could not offer.

During the fall term of 2008, one of the most significant work areas for the aftercare group was to prepare for reactions that might be brought on by the one-year anniversary. Coordinators helped plan events marking the anniversary in cooperation with school personnel and students. They arranged for and prepared personnel to assist on the anniversary. The Kauhajoki shooting, which happened six weeks before the anniversary of the Jokela shooting, understandably affected the plans. Significant trauma reactions were re-experienced because of the new tragedy, so crisis support in Jokela was increased after the Kauhajoki shooting.

### FOLLOW-UP STUDY

A school shooting is a traumatic event that has many levels of exposure. For the follow-up study at Jokela, school personnel were divided into three groups based upon their level of exposure: moderate, significant, and extreme (Lyytinen, 2010). Individuals in the moderate exposure group were not present at school on the day of the shooting; those in the significant exposure group were present but were not directly involved; and those in the extreme exposure group saw someone being shot, saw bodies, were shot at, or were otherwise directly involved.

It was hypothesized that personnel who belonged to the extreme exposure group would have more PTSD symptoms than other groups, but this could not be confirmed. Findings were not statistically significant; however, the personnel in the extreme exposure and significant exposure groups had more symptoms of PTSD than those in the moderate exposure group. Three out of four in the moderate exposure group did have some symptoms of PTSD.

Previous studies have suggested that 9–20 percent of those exposed to any traumatic circumstance will develop symptoms of PTSD (e.g., Breslau et al., 1998; Kessler, Sonnega, Bromet, Hughes, & Nelson, 1995). However, following the Jokela shooting, the occurrence of PTSD symptoms among personnel in the study was considerably higher: 87.5 percent had at least some symptoms of PTSD four months after the shooting. Almost 80 percent still had some symptoms of PTSD a year after the shooting. These findings suggest that most school personnel experience high levels of stress after a shooting and need support in dealing with the trauma (Newman et al., 2004).

Support is needed for several years after a shooting, and a decrease in services should be gradual. Previous studies have found that PTSD can cause other psychological problems (Breslau, 2002). High stress levels also have negative effects on physical health and can lead to professional burnout. These findings together with the knowledge that some personnel do not voluntarily seek support after violent incidents at their schools (Kondrasuk, Greene, Waggoner, Edwards, & Nayak-Rhodes, 2005) suggest that it is essential to actively encourage personnel to take advantage of support services.

It is important to note that the group that began progressive treatment immediately after the Jokela shooting had higher levels of trauma response at the start of the treatment, yet their PTSD symptoms significantly decreased over time. Previous research shows that if PTSD goes untreated, it can lead to other psychological disturbances, such as anxiety disorders and depression (e.g., Breslau, 2002; Khamis, 2008). It is important to begin treatment as soon as the person is psychologically ready.

## DISCUSSION

After a traumatic incident, victims benefit from opportunities to talk about their experiences, be listened to, and be understood. It is important for all involved to learn about the normal reactions to traumatic experience and to receive support in the recovery process. While trauma responses can be overwhelming, in most cases, symptoms gradually decrease with the help of social and professional support. Group counseling sessions and education on the psychology of trauma are beneficial interventions.

The Jokela follow-up study revealed that for school personnel, the re-experiencing symptoms (e.g., flashbacks, intrusive memories) increased in the year after the shooting. The increase might stem from a number of factors. As time passes from the initial event and feelings of basic safety begin to be rebuilt, the traumatized individual can start to process the event more deeply. In the beginning of the healing process, it is normal to avoid thoughts that cannot be psychologically handled. In Jokela, school personnel returned to the school soon after the incident. Together with psychologists from the FRC group, they checked the school building and were allowed to work through the fear and other feelings that the building and the event had aroused. When classes resumed, however, they and their students had to face the crime scene daily.

In previous studies, confronting the fear has been found to be helpful in promoting recovery (Carlson & Dalenberg, 2000) and might facilitate processing of normal crisis reactions. The follow-up study at Jokela found that avoidance and arousal symptoms had decreased even though the nation was faced with another school shooting. This finding might be explained by the school personnel having faced their fear when returning to the crime scene and to their daily work routine. Crisis interventions also likely contributed to the decrease in PTSD symptoms. With the exception of one staff member, all school personnel received psychological group interventions in the first four days after the shooting. These sessions provided participants with information on normal crisis response, which seemed to help them deal with the situation.

Any new threat to school communities unsettles the sense of basic safety even in schools that are not directly affected by a shooting or threat. When a nation is faced with two shootings in a short period of time, school communities everywhere feel heightened anxiety and risk. This fact suggests that it is important to pay close attention to the welfare of all school personnel after a school shooting, not just those at the attacked school. The follow-up study supports findings from a previous study on community level post-traumatic stress after school shootings, namely that new threats to schools bring up the memories, increase anxiety, and lower the feeling of security (Palinkas, Prussing, Rexnik, & Landsverk, 2004).

## CONCLUSION AND RECOMMENDATIONS FOR AFTERCARE FOR SCHOOL PERSONNEL

School personnel at Jokela School Center continued to experience PTSD symptoms over a year after the event. Intervention strategies for responding to a school shooting are most beneficial when they address both the exposure level and any physical or psychological distress. Furthermore, it must be remembered that a school shooting affects not only the attacked school but also schools in other regions. Therefore, it is important to screen and provide services to personnel at other sites as well.

School personnel in Jokela and those at the university in Kauhajoki continue to work through heightened levels of stress. Support services for school personnel and students may be necessary for a long time. The following recommendations for assisting schools in the aftermath of a traumatic event are based on interventions provided at both of these school sites.

## PREPAREDNESS FOR CRISIS SITUATIONS

- Every school should develop a plan for crisis situations, assigning the primary tasks each member of the school community will carry out in the crisis situation.
- The crisis plan should be kept up-to-date.
- The crisis plan should be rehearsed regularly.
- The crisis plan should include information about how school personnel can protect themselves against the overwhelming stimulus burden that can result (e.g., from presence of the media, ongoing news coverage, and interest from outsiders in major crisis situations).

## PLANNING AND ORGANIZATION OF AFTERCARE

- Psycho-social aftercare should be systematically directed and organized.
- Psycho-social aftercare should be coordinated by professionals who have knowledge of crisis reactions and traumatization.
- Aftercare services will be required for a long time. The financial strains this creates for the school and municipality should be taken into account in planning the services.
- In the aftermath of a crisis, strong and supportive management and a visible presence at the school are needed.
- Local resources should be utilized in aftercare when suitable. The resources should be charted and evaluated in advance.
- If persons primarily responsible for school administration (e.g., school principal) are killed or injured, they should be replaced by persons who have not been directly affected by the traumatic event but who know the networks and services related to the affected school and municipality.
- Education about common crisis reactions should be provided by health care professionals to school personnel and their immediate families as well as to students and their families.

- School personnel should visit the school premises before the work continues. It is important that the visit is guided by psychologists who can help the personnel to psychologically "re-take control" of the school building.
- The school premises should be renovated before teaching at the school continues. Any physical reminders of the shooting need to be removed.
- Returning to work may help the personnel in their recovery toward full work capabilities. Opportunities to do so should be provided, taking into account the current capabilities and strength level of the personnel. Part-time sick leave, such as shortened work days or weeks, is a possible way to support returning to work before the individual has fully recovered.
- It is important to plan and evaluate the need for support and aftercare services at both group and individual levels, and anyone who has been exposed to the traumatic event should be screened as thoroughly as possible.
- Screening should be continuous and not be limited to the immediate aftermath, as trauma reactions and the need for support vary over time as well as between individuals. Especially in case of severe exposure to a traumatic event, the shock phase may be prolonged.
- The varied exposure of individuals to the traumatic event should be considered when planning aftercare services. For example, for psychological group interventions, groups should consist of people exposed to a common stressor.
- Trauma exposure affects the body, mind, emotions, and behavior. Aftercare services should be diverse enough to take all of these factors into account.
- The support services should be easily accessible for school personnel: be convenient, require minimal bureaucracy, and be affordable.
- In the aftermath of a school shooting, rumors and threats are common. It is the responsibility of the local police to assess and evaluate the threats and respond to them accordingly. Any threats should be promptly reported to the police. To prevent spreading of rumors and to reassure the school community, a plan for communicating actual facts should be prepared.
- Cooperation with possible student welfare groups is important. They play an important role in coordinating the evaluation of psycho-social services the students and their families need.
- Psycho-social support personnel are at risk of vicarious traumatization. Support personnel who work within the community are exposed to the trauma, potentially for a long time, and should be assessed for traumatization and provided necessary counseling services.

## MEDIA AND COMMUNICATION

- The aftercare group should cooperate with local media to inform them about crisis reactions, aftercare services, and current topics related to the situation. It is important to make a communication plan that takes into consideration the phases of crisis reactions.
- Media can use their resources to provide factual information to the community and help keep rumors from spreading.

## REFERENCES

Breslau, N. (2002). Epidemiologic studies of trauma, post-traumatic stress disorder, and other psychiatric disorders. *Canadian Journal of Psychiatry, 47*, 923–929.

Breslau, N., Kessler R. C., Chilcoat, H. D., Schultz, L., R. Davis, G. C., & Andreski, P. (1998). Trauma and post-traumatic stress disorder in the community. The 1996 Detroit area survey of trauma. *Archives of General Psychiatry, 55*, 626–636.

Carlson, E. B., & Dalenberg, C. (2000). A conceptual framework for the impact of traumatic experience. *Trauma, Violence, & Abuse, 1*, 4–28.

Kela. (2011). The social insurance system of Finland. Retrieved May 24, 2011, from http://www.kela.fi/in/internet/english.nsf

Kessler, R. C., Sonnega, A., Bromet, E., Hughes, M., & Nelson, C. B. (1995). Post-traumatic stress disorder in the National Comorbidity Survey. *Archives of General Psychiatry, 52*, 1048–1060.

Khamis, V. (2008). Post-traumatic stress and psychiatric disorder in Palestinian adolescents following intifada-related injuries. *Social Science and Medicine, 67*, 1199–1207.

Kondrasuk, J. N., Greene, T., Waggoner J., Edwards, K., & Nayak-Rhodes, K. (2005). Violence affecting school employees. *Education, 125*, 638–647.

Lyytinen N. (2010). Post-traumatic stress symptoms among school personnel after the Jokela school shooting: A longitudinal study of exposure, interventions and symptom changes. Retrieved November 15, 2011 from https://oa.doria.fi/handle/10024/61765.

Newman, K. S., Fox, C., Harding, D. J., Mehta, J., & Roth, W. (2004). *Rampage: The social roots of school shootings*. New York: Basic Books.

Palinkas, L. A., Prussing, E., Rexnik, V. M., & Landsverk, J. A. (2004). The San Diego East County School shootings: A qualitative study of community level post-traumatic stress. *Prehospital and Disaster Medicine, 19*, 113–121.

Saari, S. (2005). *A bolt from the blue: Coping with disasters and acute traumas*. London: Jessica Kingsley Pub.

Sarason, I. W., Levine, H. M., Basham, R. B., & Sarason, B. R. (1983). Assessing social support: The social support questionnaire. *Journal of Personality and Social Psychology, 44*(1), 127–139.

Shapiro, F. (1989). Efficacy of the eye movement desensitization procedure in the treatment of traumatic memories. *Journal of Traumatic Stress, 2,* 199–223.

Shapiro, F. (1995). *Eye movement desensitization and reprocessing: Basic principles, protocols and procedures.* New York: Guilford.

\* \* \*

## CONTRIBUTOR NOTES

Authors **Kirsti Palonen** and **Nina Lyytinen** worked as part of the team of psychologists and crisis workers hired by the municipality of Tuusula after the Jokela school shooting to coordinate the psycho-social aftercare through May 2009. Psychologist/ psychotherapist Kirsti Palonen, who was the leader of the aftercare coordinators group, has a long history of working with crisis situations and catastrophes since the year 1990. Psychologist Nina Lyytinen worked as part of this group and completed a follow-up study of exposure, interventions, and symptom changes among school personnel after the Jokela school shooting.

---

### WHAT'S NEXT?

After a traumatic event, schools and universities face an increased need to respond to the mental health concerns that result. In Chapter 10, Russell T. Jones and several of his doctoral students share the model for mental health response and intervention that was employed after the shootings on their campus, Virginia Tech. Educators, while not expected to be therapists, should be aware of what will be involved in the way of mental health support so they can factor this essential component into their institutional recovery plans.

# CHAPTER 10

# TRAGEDY AT VIRGINIA TECH: RECOMMENDATIONS FOR MENTAL HEALTH RESPONSE TO CRISIS ON CAMPUS

RUSSELL T. JONES, KATHARINE DONLON, KELLY DUGAN BURNS, KATHRYN SCHWARTZ-GOEL & MARY KATE LAW

## BACKGROUND: THE EVENT

On the morning of Monday, April 16, 2007, Virginia Tech (VT) student Sueng Hui Cho shot and killed 2 students in a campus dormitory, West Ambler Johnston Hall. In a separate incident, approximately two hours later, he killed 30 students and faculty members in Norris Hall, an academic building. Cho wounded seventeen other students and faculty members in the attack and then committed suicide. In addition to those who were physically injured, many more were affected either by witnessing the events or because they knew someone directly impacted (*Virginia Tech Review Panel*, 2007).

Shortly after the first attack, an e-mail was sent to all VT students, faculty, and staff informing them of the shootings. As the second attack was ending, another e-mail was sent to alert everyone of a possible gunman on the loose, and the same message was delivered over loudspeakers around campus. Following the Norris Hall attack, several campus buildings were locked down for hours. Multiple e-mails were sent during the lock-down period, instructing people to stay where they were, informing them of multiple victims, and updating them that the shooter had been captured. Another e-mail was sent to notify people that classes had been cancelled for the remainder of the day.

Rescue squads, ambulances, and police SWAT teams responded to the incidents. Additionally, mental health recovery efforts were initiated quickly after each attack. Within 30 minutes of the first attack, counseling psychologists from Cook Counseling Center on campus began crisis counseling with friends of the victims and fellow dormitory residents. After the second shooting, community mental health providers from Blacksburg and the campus disaster response network responded immediately to assist the families directly impacted by the shootings. These providers gave support as families received death notifications and as they waited for the release of their loved one's body. The day after the shootings, a network of professionals met to begin planning the ongoing mental health response to the tragedy. This planning network included members of the Cook Counseling Center; representatives from the clinical psychology, counselor education, and marriage and family programs; the employee assistance program; community agencies; and the local mental health association.

In the week following the shootings, psychologists and counselors held more than 120 group sessions, ranging in size from a few people to thousands who attended the convocation on the day after the attacks. Psychologists from the counseling center provided special assistance for those students who lived in the dormitory, anyone in Norris Hall at the time of the shootings, roommates of the deceased students, faculty and students in classes attended by the victims, and others who knew the deceased through campus organizations. More than a thousand students and faculty received individual counseling at multiple sites around campus in the week following the attacks. When classes resumed one week later, volunteer counselors circulated to classes where any of the injured or deceased had been a student. Many faculty, students, and student groups received presentations on trauma, post-traumatic stress, and wellness. Information about resiliency following trauma was prepared and disseminated, both in person and through the Internet.

VT's medical center sent personnel to the hospitals where the injured were being cared for to assure them of ongoing access to services. The Cranwell International Center, which serves international students on campus, established personal connections with each of the international students, and provided support to the families and friends of many of the victims who had language or cultural needs. Staff members and volunteers called the rest of the international undergraduate students and then called the remaining international graduate students (K. Beisecker, personal communication, April 8, 2011).

## EXPERIENCES AND RESPONSES

PROFESSOR RUSSELL T. JONES

What started as a typical Monday morning turned out to be one of the most tragic days in the nation's history. As the events of this day began to unfold, I was in the process of packing for one of my many trips to the Gulf Coast with the goal of teaching disaster preparedness skills to kindergarten children. We were about to embark on our first initiative to help children learn what to do in case of emergencies in that area of the country. As I packed, I could hear the NBC news broadcast in the background. As I recall, I heard something like the following statement: "Two students have been shot at Virginia Tech." I immediately turned to the television in disbelief. As I stayed glued to the TV, the number of individuals reported as being shot began to steadily increase.

The call to respond to yet another disaster situation was not new. My 30-plus years of research and clinical expertise in the areas of emergency functioning and responding to natural and technological disasters had well prepared me to deal with crisis. My response to residential fires in several states, wildfires in California and Florida, and Hurricanes Andrew, Katrina, Rita, and Ivan had proven to be valuable learning experiences. Despite these efforts, I was not fully prepared to deal with the VT tragedy. The difference was that this disaster was unfolding in "my own backyard." My students, fellow faculty members, and staff were all vulnerable.

During the acute phase of a crisis, our immediate attention is often focused on those we are closest to who may be in harm's way. Therefore, my thoughts immediately turned toward my wife who was at a routine doctor's appointment near campus. The phrase "a conflict of loyalties" consumed my thinking. I wanted to make sure that she returned safely, but at the same time, I wanted to rush to campus. Within a short period of time, she was safely home. Thank God!

While hastily leaving my bedroom and glancing at my half-packed suitcase, I was reminded of a statement made by a teacher in the Gulf Coast shortly following Hurricane Katrina. Standing in front of her partially destroyed elementary school, she commented on the horrific impact of the storm on her home, followed by a statement of its impact on the children: "I have a hole in my roof, but a greater hole in my heart because no one is looking out for the kids." I immediately identified with these words in a way that I never had before. For the first time, that statement now referred to the "kids" at VT. I needed to know firsthand about any students, faculty, or staff who had been hurt or injured by these shootings. Off to campus I went.

Many of the initial phone calls I received were from graduate students in the Psychology Department, and more specifically members of our R. E. A. A. C. T. (Recovery Efforts After Adult and Child Trauma) team. They expressed their desire to put into practice what they had learned regarding crisis responding. Five of my graduate students immediately became involved in initial recovery efforts. What I found most encouraging was the fact that they did not wait to ask me what to do, but moved forward in partnering with mental health professionals at the university and the community to promote structure, safety, and comfort to individuals in need. While several worked with members of the American Red Cross, the rest worked with other local organizations. A meeting with the entire Psychology Department was called, with the goal of developing strategies to assist in the acute and intermediate aftermath of the shootings. Discussions during this meeting provided our students and faculty with guidance for moving forward.

My immediate goal during the initial phase of the acute recovery process was to meet with first responders and administrators, and with victimized students, faculty, and staff. As I arrived on campus, I was pleased to see the number of first responders but shocked to see the enormous media presence. While the task of "getting the story out" is of utmost concern following any traumatic event, I was quite apprehensive about how the media would do this. Headlines, stories, and sound bytes at the expense of the VT community were quite unsettling.

Unlike the Gulf Coast where the mental health infrastructure was wanting at best because of fragmentation and lack of resources, our community was just the opposite. As I made my way into the command center, I saw representatives from the New River Valley Community Health Center, Cook Counseling Center, the University's Psychological Services Center, the New River Valley Community Services Board, the local Red Cross, and many others. The benefits of having worked together with these organizations over the past 20 years became evident immediately.

Over the next few days, in numerous meetings, e-mails, and phone calls with community partners, one of the most important tasks was to prepare students and faculty to resume classes. It was decided that pairs of mental health professionals would attend each class and share information with students and faculty. Psycho-education was provided to inform individuals of "normal reactions" to this "abnormal situation," ways of coping, as well as the location of sites where assistance could be obtained. Handouts from the American Red Cross, the American Psychological Association, the National

Child Traumatic Stress Network, and the American Psychiatric Association were modified to address the present situation and made available. As a result of this collective effort, we felt that many of the immediate fears and reactions of students, faculty, and staff were lessened.

Shortly after the shootings, a meeting was arranged with our University Provost, Dr. Mark McNamee. Issues related to safety, assessment, intervention, and funding were discussed. Among the initiatives resulting from this and later meetings were efforts to obtain funds to assist us in moving forward with the recovery phase. I became a member of a five-person working group tasked with developing a grant proposal for the U.S. Department of Education. Initial foci of this endeavor were mental health recovery and threat assessments. After hundreds of e-mails and countless hours of meetings, a final proposal was submitted and funded by the Department. A similar endeavor initiated by members of the community brought in funds from the U.S. Substance Abuse and Mental Health Services Administration (SAMHSA).

## KELLY DUGAN BURNS

That morning, I was attending a class near Norris Hall. A student who arrived late to class was frantic with the news that someone had been murdered on campus. From then on, we checked the VT website for updates on what was occurring. The updates continued, and soon we heard police on their bullhorns instructing us to stay away from windows and to remain where we were. Fear quickly ensued as the updates revealed that the number of possible deaths continued to rise, and that one or several murderers were still on the loose. Later that day, we and the nation would find out the devastation that our campus experienced and the total number of lives lost. As a budding trauma researcher and clinician, this changed the way I thought about "trauma."

Like many in my department, I volunteered with the American Red Cross to offer my assistance. Several went to the hospital to be with the families of those who were wounded. Others went to the Inn at Virginia Tech, which served as a gathering spot for families, to try and console the families who had lost a loved one. Along with three other clinical psychology graduate students, I was asked to work with the FBI's Office of Victim's Assistance. Our job as mental health workers was to assist students and faculty in going back into Norris Hall several days after the shooting. We aided them in collecting personal belongings they had abandoned on that fateful morning and provided any immediate support needed.

I began teaching VT undergraduates in the fall 2009 semester, more than two years after the shootings. Despite the length of time that had passed, and the fact that several of my students were not present on campus when the shootings occurred, the experience had considerable influence on my instructional approach. While the content of the material remained unaffected, I worked hard to get to know my students individually and remained open and available to each student's needs related to his or her ability to learn. For instance, if students had family emergencies, mental health difficulties, or other problems that caused them to miss graded assignments, I would work with them to determine a plan of action that would suit both the student's and the course's needs, rather than simply referring to a course tenet stated in the syllabus.

## KATHRYN SCHWARTZ-GOEL

The morning of April 16th I was in class near Norris Hall when I received an e-mail stating a gunman was on campus and to stay away from windows—nothing else. The e-mail did not appear to come from a VT e-mail account, so I dismissed it as weird. Eventually another classmate said that she had received the same e-mail. That's when we began to worry. Around the same time we started hearing and seeing ambulances zooming by the large windows in our classroom. Still uncertain about what was going on, we checked the websites of the local news and CNN and got the first clear news that there was indeed a gunman on campus; however, no one was sure exactly what had happened.

Although the instructor attempted to continue class, everyone's attention was elsewhere. We all called our parents to let them know we were safe, and then ran to our cars to leave campus, despite the lockdown. The rest of the day was spent glued to the TV, as more and more news came in about the devastation and horror that had taken place. I attended the candlelight vigil that night with many friends and spent the next few days together processing what had happened.

We also attempted to help the Red Cross with their efforts, convening first at their headquarters and then at the local hospital. Initially, our help was not needed, as everyone was trying to organize their efforts. I ended up leaving town for the weekend to get away. My husband and I went to Charlottesville and were amazed by the support for VT there. Everywhere we looked, people were wearing VT paraphernalia, despite being the home of rival University of Virginia. Many businesses had signs up supporting VT and even discounts for students.

In addition to the personal effect the shooting had on me, it also affected my activities and progress in my program. It was difficult to concentrate immediately afterwards. Along with my entire cohort, I ended up 6 months to a year behind in the program as a result of the shootings.

There are some things that would have been helpful to have in place. For example, it may have been helpful for *all* professors to provide a quick overview of the effects of trauma and campus resources that were available. If professors did not feel equipped to do this, mental health volunteers could have fulfilled this need. I am not sure of all the community resources that were employed, but it may also have been helpful to educate local students and community members as well.

## MARY KATE LAW

The shootings occurred during my first year of graduate school. I have often wondered if the consuming nature of graduate school increased my inability to imagine such an event taking place, especially at the university I attended; however, it seems that the nation as a whole was shocked and in disbelief as well. The first few days following the shootings were surreal. Phone calls, text messages, and e-mails poured in from family and friends, as well as from individuals I had not seen or spoken to in years. I carefully read through VT press releases to assess who had been injured or killed and if I knew any of them. I selfishly was grateful that I did not. It was also a time to gather with other Hokies and know that the world had not stopped turning. The heartwarming broadcasts, the president's visit, and the understanding faculty were all sources of comfort.

Finding meaning in April 16th has been a core component of my and others' healing. It was incomprehensible for such events to take place without something positive evolving. For example, the Virginia Tech Center for Peace Studies and Violence Prevention was established—a perfect example of survivors coming together for a greater purpose. Other positive outcomes spurred by this tragedy include a broader awareness of violence and warning signs among students, as well as tracking of mental health referrals. Threatening communication and troubling behaviors are taken very seriously, and referrals are followed more carefully to ensure that services are received. I believe these efforts will be influential both in preventing future violent acts and in improving the lives of struggling individuals.

## RESPONDING TO MENTAL HEALTH NEEDS: A CONCEPTUAL FRAMEWORK

Viewing VT's response to the mental health needs brought on by this tragedy through a conceptual lens, it might be useful to trace actions taken during the immediate aftermath of the shootings. Robert Pynoos at UCLA and Shep Kellman at Johns Hopkins provided valuable input early on. Based on Pynoos's previous work with school shootings, Shep Kellman's research in the area of prevention/intervention, and the first author's clinical and research efforts with natural and technological disasters, the following working outline and objectives were developed.

*Stage 1:* Public Health/Prevention Framework

- Articulate a shared vision

*Stage 2:* Coordination and Development of Services

- Coordinate existing services
- Identify gaps in existing services
- Develop new services
- Implement prevention model (Universal/Selective/Indicative)

*Stage 3:* Coordination and Maintenance of Public Health/Prevention Strategies at the University and Surrounding Community

- Continue to obtain input from university students/faculty/staff and members of the surrounding community
- Continue to obtain advice of the scientific advisory board

Of particular relevance for intervention efforts in the immediate aftermath of the shootings was the examination of the UCLA Trauma Psychiatric Program's three-tier model for schools by Pynoos and his colleagues. This mental-health-based model was employed after the shootings at Santana High School (Pynoos, Goenjian, & Steinberg, 1995) and at Columbine High School (Weintraub, Hall, & Pynoos, 2001). Because of my previous interaction with Dr. Pynoos, I invited him to campus following the shootings.

It should be noted that many of the steps taken by VT were aligned with this model. That is, many actions of the psychologists, counselors, graduate students, employee assistance program, community agencies, local mental health association, as well as VT administrators, faculty, and staff were consistent with the various tiers of this model. Examples of efforts falling under each tier are presented below.

*First tier:* Early intervention by mental health professionals

First-tier actions at VT included the response to the shootings within 30 minutes and the efforts to enhance safety and support, restore normalcy, and provide crisis intervention. In addition, the university focused on meeting student needs on the first day of classes after the shootings, and a network of professionals provided immediate, large-scale support throughout the campus. Approximately three hundred volunteers visited classes that any of the injured or deceased had attended to provide psycho-education regarding reactions to trauma as well as information about where mental health assistance could be obtained.

*Second tier*: Implementation of specialized mental health services for those with severe persistent distress and the assessment of individuals' needs for mental health services

After the initial stage, it was important to assess individuals' needs for mental health services, coordinate existing services, identify any gaps in services, and develop new services to meet identified needs. Across the VT campus, assessment of needs was initiated, and specialized mental health services for those with severe persistent distress were implemented. VT offered greater availability of mental health services, in part, by the campus counseling center temporarily establishing "walk in hours" and extending normal hours of operation (7:00 a.m. to 9:00 p.m., including weekends). Three additional sites on campus were made available where faculty and staff could obtain services.

A small team guided by Dr. Ron Kessler at Harvard University developed a needs assessment survey to ascertain access to treatment and inform the university about students, faculty, and staff needs. In addition to being useful for guiding intervention efforts for those impacted by the shootings, it served as a means of determining the effect of this trauma on an individual's functioning.

More than four thousand faculty, staff, and students responded to the survey. Of the student respondents, 10 percent indicated that they had received counseling following the shootings, and 28 percent of the students reported that they intended to seek counseling (Hughes et al., 2011). In the immediate aftermath of the shootings, usage of the campus counseling center increased by 35 percent (Flynn & Heitzmann, 2008).

*Third tier*: Provision of assistance for those in need of more intensive services

The third tier of this model involves obtaining help for those who need more intensive services than were provided in Tier 2. That is, special care was made available for those who lived in West Ambler Johnston Hall, as well as for those who were in Norris Hall during the time of the shootings. Individuals who were taken to local hospitals for immediate attention fall into this tier. VT's health center was also involved in this process, while local and regional Emergency Medical Services personnel provided Critical Incident Stress Management activities to individuals affected by the shootings.

## WAYS EDUCATORS CAN SUPPORT RECOVERY

These mental health services have implications for educational administrators as well as for mental health counselors. In addition, other aspects of mental health awareness have significance for classroom teachers and faculty as well. Notably, the five components described by Hobfoll and colleagues (2007) as crucial for individuals who have recently experienced a mass trauma event are instructive for all who are in the educational environment. Whether in a classroom, on the campus, or at a community venue, the following are essential elements in the aftermath of trauma.

1. **Promote a sense of safety:** Because of the unexpected nature of trauma events and the disruptive impact that they can have on people's lives, creating a sense of safety for affected individuals after the event has a positive effect on the development of and subsequent duration of potential traumatic stress symptoms. Students cannot learn and faculty cannot teach if they are anxious or fearful.

2. **Promote calming:** It is helpful to encourage *calming* after people have experienced traumatic events, in order to reduce the strong level of emotion likely to be experienced. When students or faculty appear to be lost in emotion or panic, it helps to provide a compassionate, grounded presence.

3. **Promote a sense of self- and collective-efficacy:** To counter the feelings of helplessness and loss of control that are experienced in a traumatic situation, it is important for individuals to regain a sense of control over future events and challenges. It is also necessary that individuals and groups believe that they can experience positive outcomes.

4. **Promote connectedness:** A sense of connection among individuals as sources of social support is remarkably effective in alleviating stress and other negative reactions to the trauma. Schools and universities can help foster such connectedness by providing opportunities to get together with others through collective campus functions as well as through online social networks.

5. **Promote hope:** Optimism is associated with positive outcomes and is a powerful tool for reducing the likelihood of negative psychological effects following a trauma. It is particularly important because of the "shattered worldview" that often accompanies traumatic events.

## LEARNING FROM THE TRAGEDY AT VIRGINIA TECH

### RECOMMENDATIONS FROM THE GOVERNOR'S PANEL

The tragedy at VT forced the Commonwealth of Virginia to take a critical look at policies and procedures on both state and university levels. The findings from the resulting report will allow colleges and universities across the country and the world to enhance crisis prevention and response. Several recommendations applicable to reclaiming a school or campus in the aftermath are summarized below. (The complete report is available at http://www.governor.virginia.gov/TempContent/techPanelReport.cfm.)

**Security** is an essential focus, both for preventing mass trauma events and for responding to crises that do develop. In particular, the report suggested that universities carefully consider their emergency planning, conduct a risk or threat assessment, and determine an appropriate level of security for the campus. Additionally, universities need to update and enhance their Emergency Response Plans to ensure they are in compliance with state and federal guidelines, have a threat assessment team, and maintain their campus alerting system so that it operates in the most effective and efficient way possible. In the event of an emergency, the campus community needs to be informed immediately, with clear instructions about the nature of the emergency and subsequent actions to be taken. Campus police and administrative officials need to be able to send such emergency messages. Campus police should be trained to respond to active shooters, and the chief of the campus police department should be a member of the threat assessment team.

**State mental health policies** should be reviewed to ensure that information concerning a student who may pose a risk is acted on appropriately. For example, if a student needs to be detained for mental health concerns, the time allowed for detention must be sufficient to allow for an in-depth mental health assessment. Moreover, emergency physicians should be permitted to issue temporary detentions based on their evaluations. Finally, a temporary detention hearing for students with serious mental health concerns, held at the university, should require that students present the report from an emergency physician's evaluation, the results of lab or toxicology testing, reports of previous psychiatric history, and admission notes.

**Privacy laws** need to incorporate a "safe harbor" specification that protects anyone who discloses information about an individual with a potential mental health concern and that protects the identified person, other individuals, or the community at large. Additionally, federal regulations governing the sharing of student information (i.e., FERPA) should make an exception to their mandate that this type of mental health record cannot be shared without a person's consent. In particular, treatment records from university clinics should be disclosed without student consent and be made available to any health care provider. The report suggested that the National Higher Education Association should create guidelines and training on the issue of information sharing.

**Laws governing the purchase of guns** and campus policies for possession of firearms on campus need to be assessed. The report indicated that the Commonwealth of Virginia should mandate mental health background checks for all firearm sales and recommended that all individuals who are determined, by a hearing, to be a danger to themselves should be entered into the Central Criminal Records Exchange database. This should occur whether or not they voluntarily agree to receive treatment. Furthermore, the report indicated that the ability of universities and colleges to ban guns on campus should be determined by the attorney general. Universities and colleges should also clearly state their policies regarding guns on campus.

**Crisis communications** protocols should ensure that initial alerting messages and subsequent crisis information are circulated as soon as danger is recognized. Moreover, the members of the university community who receive a message should be encouraged to inform others of the situation. Additionally, the report suggested that universities have multiple methods for transmitting information, several of which are not dependent on technology. Finally, plans for canceling classes or closing the campus should be detailed in the university's emergency operations plan.

**Active Shooter Protocols** should be the subject of campus and community police training. Additionally, it is important for dispatchers to use caution when providing instructions to individuals in an emergency situation without knowing many details about the situation. Universities and colleges should check the hardware on exterior doors to ensure that they are not subject to being chained shut. Schools are urged to take bomb threats seriously and to encourage community members to report them immediately.

IMPLEMENTING RECOMMENDATIONS AT VT

While many steps have been taken in response to the report of the Governor's Panel investigation, those that appear to have led to particularly meaningful outcomes include actions consistent with the three-tier model by Pynoos and colleagues previously discussed. Some of the changes that have been made since the shootings, articulated by the Associate Vice President of University Relations, Mr. Larry Hincker, are summarized as follows.

Several changes were made following the April 16 tragedy, including establishing new offices, security updates, and staffing improvements. The Office of Recovery and Support was established to serve as a liaison between the university and the families of students who were injured or killed. Although this office now has just two full-time employees, at its peak it had three counselors, a case manager dedicated to the academic needs of the injured students, and several case managers who dealt with the specific needs of each family. The majority of the *Hokie Spirit Fund,* totaling $10.5 million that was spontaneously donated by a variety of sources to help with the recovery, was divided among the families of the injured or deceased individuals.

Several changes regarding campus security were enacted shortly after the shootings. Specifically, 11 new positions were added in the campus police department. The university mandated that every residence hall be locked and accessible only to those students who reside therein. Furthermore, locks were installed on all classroom doors, and crash bars were placed on all doors to prevent chaining (as the gunman had done in Norris Hall). Additionally, new campus sirens and speakers were added. The Emergency Notification System, a web portal, was created to link at least six methods of communication (i.e., mobile devices, electronic signs, computer alerts, blast e-mail, social networking blasts, and website news), and the ability to send alerts via text messaging was established. Finally, the university expanded the number of people with the authority to send messages.

A number of staffing changes were also enacted. The budget for the Cook Counseling Center was increased by 50 percent, and six full-time counselor positions were added. A case manager position was added at the Dean of Students Office.

A Threat Assessment Team was created after the shootings. This team, headed by the Chief of the VT Police Department, is composed of individuals from several different areas, including the Legal Team, the Police Department, the Dean of Students, Cook Counseling Center, Residential and Dining Programs, and Judicial Affairs.

VT recognized the need for remembrance following the tragedy. A memorial was created on the Drillfield, which is the center of campus, to remember the individuals who were killed. Additionally, an archive was established to hold the thousands of items and expressions of support that were sent from around the world following the event (L. Hincker, personal communication, February 23, 2011).

### THE VT DISASTER MENTAL HEALTH GUIDE

In the event of a future, potentially tragic event, it is vital to have a plan in place to respond to mental health emergencies. Since the shootings, VT has been actively involved in the development of a plan to assist with future events. What follows is a list of key recommendations from the VT Disaster Mental Health Field Operations Guide that may be of benefit to other colleges and universities.

I. Recognize that a college or university is a community within a community.
   a. Take into account that a disaster not only affects the university campus but also the community as a whole. Situations of this

magnitude call for external support from the community, region, and possibly state and federal sources.

   b. Provide workshops on disaster mental health free and open to members of the campus community and the local community. Participants in these workshops can become trained volunteers to help out in the event of a disaster.

   c. Determine procedures for initiating disaster mental health response (e.g., by the police/fire/rescue, or by the university office of emergency management).

  II. Designate possible sites for an assistance center(s) where individuals can go for information, support, and connecting with resources.

 III. Require appropriate identification (e.g., badges) to participate as a university support staff or mental health professional in the event of a disaster. This can help distinguish trained professionals from people posing as trained professionals (e.g., reporters).

 IV. Identify a media relations team.

## CULTURALLY COMPETENT CARE

A key component of any recovery effort should be culturally competent care. A university community is likely composed of individuals representing a variety of races and ethnicities; thus consideration needs to be given to cultural sensitivity and to potentially greater needs among certain groups on campus. At VT, for example, it was reasoned that special attention should be rendered to the needs and perceptions of Korean students, given that the shooter was of Korean heritage. Many actions by the Cranwell International Center and other university offices assisted Korean students, faculty, and Korean-American communities.

To meet these needs, guidelines from a variety of resources were reviewed and adapted. One example was the discussion of issues related to stigma with "people of color." Guidelines from a model (Jones, Hadder, Carvajal, Chapman, & Alexander, 2006) for attending to needs of people of color were adapted to address several challenges faced by those of Korean heritage. This model addresses three important areas, namely, mistrust/beliefs, barriers to access, and culture/linguistics. Each is essential for the successful conceptualization, assessment, and treatment of problem behaviors by a person of color. A brief outline of each area of the model follows.

### MISTRUST/BELIEFS

Given the stigma associated with mental health assessment and treatment under the best of circumstances, special attention is warranted when working with members of minority and marginalized communities. Mistrust is a major obstacle for participation by people of color in treatment and/or research-related activities. Recovery efforts include the following considerations:

- Discuss and assess, when possible, levels of mistrust.
- Locate community gatekeepers and ask their input and involvement.
- Develop relationships with leaders and members of the community.
- Establish rapport by building bonds with members of the community.
- Involve representatives from affected groups to assist in all phases of activities.
- Engage people of color and individuals from marginalized communities as role models.
- Develop an understanding and appreciation of current needs and realities of impacted individuals and communities.
- Identify, appreciate, and respect differing cultural beliefs and practices.

## BARRIERS TO ACCESS

Offering incentives may increase participation of minorities, who may be unfamiliar with intervention programs and/or research projects that seek to learn from the disaster. Consideration of the following salient points greatly reduces hindrances to participation in therapeutic or research processes:

- Identify service-delivery sites convenient to the communities.
- Engage publicity campaigns likely to reach minorities.
- Provide door-to-door recruitment whenever possible.
- Always provide convenient hours of operation.
- Identify transportation services available to those involved in treatment/research or reimburse them for transportation costs.
- When possible, provide financial assistance, fee waivers, and other incentives.

## CULTURE/LINGUISTICS

Awareness, understanding, and appreciation of cultural nuances are essential if one genuinely desires to engage individuals from various cultural groups. Integrating customs, beliefs, and values of marginalized, underserved groups will go a long way to gain involvement and to widen clinicians' and researchers' perspectives. A significant barrier for non-English speaking peoples is language. Bilingual mental health professionals who are familiar with local idiomatic expressions, symbols, and concepts shared by cultural groups have been quite beneficial. Recruitment of such counseling and support professionals could do much to resolve historical barriers to participation. The following steps should be considered whenever working with minority groups.

• Plans outlining goals, policies, and systems of accountability should be developed, implemented, and assessed when engaging culturally and linguistically appropriate services.

• All team members in culturally and linguistically appropriate service delivery and research methods should be trained using culturally competent curricula.

• Translate and interpret materials and measures for participants who do not know or are uncomfortable with the English language.

## RECOMMENDATIONS FROM THE DISASTER MENTAL HEALTH SUBCOMMITTEE

The Disaster Mental Health Subcommittee of the National Biodefense Science Board, which is made up of subject matter experts including the first author, have used their collective experience and available empirical evidence to develop eight recommendations that can be used to guide mental health responses in the event of a future disaster or emergency (2010). The recommendations are grouped into three major areas: intervention, education and training, and communication and messaging.

While all three areas are important for promoting recovery from disaster, the second area, education and training, has particular significance for educators. The subcommittee's complete report of recommendations is available at http://www.phe.gov/preparedness/legal/boards/nbsb/meetings/documents/dmhreport1010.pdf.

1. **Intervention recommendations** are centered on the inclusion of the mental health response in the broader response to disasters and emergency situations. Specifically, it is recommended that the mental and behavioral health response be incorporated with public health and medical preparedness and response to a disaster or emergency event. The committee encouraged research within the disaster mental health and behavioral health umbrella. Additionally, the mental and behavioral health response to emergencies should place more emphasis on the assessment of mental and behavioral health needs.

2. **Recommendations for education and training** promote the development of skills in disaster mental health and behavioral health professionals and to encourage resilience in the broader population. In particular, disaster mental health and behavioral health training for professionals as well as paraprofessionals should be augmented. Mental health professionals should promote resilience in the population by integrating, disseminating, and continually evaluating this intervention. Finally, it is important to meet the unique needs of at-risk individuals during the disaster response process, and to ensure cultural competency.

3. **Strategies to improve communication and messaging** point to the importance of developing a disaster mental and behavioral health response plan for communication prior to the occurrence of a disaster or emergency. The response protocol needs to include a variety of communication strategies, including via the Internet.

### FINAL OBSERVATIONS FROM THE AUTHORS

RUSSELL T. JONES

To provide a context for several of my recommendations for responding to a crisis on campus, I want to share an experience that resulted in a metaphor I use to characterize post-disaster actions. Immediately following a deployment to the Gulf Coast after Hurricane Katrina, I found myself on a plane discussing my experiences as a trauma psychologist with an individual in the seat next to me. During our conversation, I learned that he was a "rocket scientist." An interesting discourse ensued.

Being completely exhausted and somewhat frustrated with the fact that the pilot had announced that we were having engine problems, I asked this rocket scientist, "What is it about a rocket engine that makes it as efficient as it is?"

He replied, "In a rocket, there is not a single engine, but multiple engines." He went on to say, "There can be three, six, or even nine engines, and what these engines do is interact. They compensate for one another." He stated that the technical name for such interaction is *gimble*. He said that when this process of *gimbling* took place, maximal thrust, efficiency, and direction resulted.

I enthusiastically replied: "What a metaphor for mental health professionals in times of crisis!"

Instances of *gimbling* were demonstrated in the actions taken by graduate students during the immediate aftermath of the VT shootings, in the university's interaction with community partners, in the collaboration among university administrators and departments, in the development of a survey to assess mental health needs, and in the efforts to raise funds to support the recovery. Indeed, as we sought to reclaim our university after the tragedy, it seems that we all became adept gimblers.

That is my advice to others who face similar difficult situations. Find ways to work together, be alert to the needs of others, identify and access resources that can be put to good use, and above all, work together for the good of all.

## GRADUATE STUDENTS

The following is a list of recommendations from the doctoral-student authors regarding changes they felt would be beneficial, not only in responding to a tragedy but also in preventing such events and in strengthening the fabric of the campus environment.

1. Continue with school activities, classes, and other functions. Allow students time to grieve but also provide a normal routine as much as possible.
2. Publicize updated security measures (e.g., increased police presence, campus-wide alert systems).
3. Enact and enforce strict zero-tolerance policies regarding violence, discrimination, intimidation, and other forms of abuse.
4. Continue efforts to build community within the university (host events, vigils, concerts, and other opportunities to bring the campus together).
5. Encourage activities to build relationships between the local community and the campus community (e.g., fairs, music festivals, picnics) that will engage students, faculty, staff, and residents of the town.
6. Provide information about effects of trauma in classrooms to help students understand that how they are feeling or reacting is normal. Provide access to mental health services.
7. Increase security presence (e.g., officers, check points, lighted emergency boxes around campus, well-lit parking lots). Evaluate campus security, and if needed, add programs (e.g., Safe Ride Home).
8. Maintain locked access to all dorms.
9. Provide special programming/workshops on violence prevention, gun laws, warning signs of violent/ill individuals, rape-prevention programs, self-defense, and related concerns.
10. Develop activist outlets: Create student-led groups such as neighborhood/campus watch; mental health outreach; and other opportunities to be involved, make a difference, and help others.

11. Demonstrate timely, supportive, committed response from university leaders (e.g., the president, board of directors, senior faculty) that conveys that they understand the gravity of the situation, are supporting the campus/community, and are working diligently to make changes so that the event(s) will not happen again or can be mitigated.

## CONCLUSION

Much can be learned from the tragedy at Virginia Tech. The process for returning a campus to functionality requires ensuring mental health needs are addressed; that all feel safe in their environment; that they are supported in their healing by a compassionate and understanding university community; and that they take advantage of the healing aspects of social networks of family and friends. It also helps when students, faculty, and staff become engaged in efforts to help others, participate in activities that connect to the larger community, and find a way toward constructing meaning in the tragedy and reconciling to the experience. With attention to lessons learned from this and other tragedies, foresight in planning for responses, developing a culture of caring across the campus community, and building strong working relationships with agencies and individuals in the local community, a college or university can transcend loss and become stronger in the aftermath.

## REFERENCES

Flynn, C., & Heitzmann, D. (2008). Tragedy at VT: Trauma and its aftermath. *The Counseling Psychologist, 36*(3), 479–489.

Hobfoll, S. E., Watson, P., Bell, C. C., Bryant, R. A., Brymer, M. J., Freidman, M. J., & Ursano, R. J. (2007). Five essential elements of immediate and mid-term mass trauma intervention: Empirical evidence. *Psychiatry, 70*(4), 283–315.

Hughes, M., Brymer, M., Chiu, W. T., Fairbank, J. A., Jones, R. T., Pynoos, R. S., Rothwell, V., Steinberg, A. M., & Kessler, R. C. (2011, July 18). Posttraumatic stress among students after the shootings at Virginia Tech. *Psychological Trauma: Theory, Research, Practice, and Policy.* Advance online publication. doi: 10.1037/a0024565.

Jones, R. T., Hadder, J., Carvajal, F., Chapman, S., Alexander, A. (2006). Conducting research in diverse, minority, and marginalized communities. In F. Norris, S. Galea, M. Friedman, & P. Watson, (Eds.), *Research methods for studying mental health after disasters and terrorism* (pp. 265–277). New York: Guilford Press.

National Biodefense Science Board. (2010). Integration of mental and behavioral health in federal disaster preparedness, response, and recovery: Assessment and recommendations. Retrieved May 20, 2011, from http://www.phe.gov/preparedness/legal/boards/nbsb/meetings/documents/dmhreport1010.pdf.

Pynoos, R. S., Goenjian, A., & Steinberg, A. M. (1995). Strategies of disaster inter-
vention for children and adolescents. In S. E. Hobfoll & M. W. deVries (Eds.),
*Extreme stress and communities: Impact and intervention* (pp. 445–471). New
York: US Kluwer Academic/Plenum Publishers.

Weintraub, P., Hall, H. L., & Pynoos, R. S. (2001). Columbine High School shoot-
ings: Community response. In M. Shafii & S. L. Shaffi (Eds.), *School violence:
Assessment, management, prevention* (pp. 129–161). Washington, DC: American
Psychiatric Publishing.

Virginia Tech Review Panel. (2007). *Mass shootings at VT, 2007: Report of the
Review Panel.* Retrieved July 15, 2011, from http://www.governor.virginia.gov/
TempContent/techPanelReport.cfm.

\*   \*   \*

## CONTRIBUTOR NOTES

**Russell T. Jones, Ph.D.**, professor of psychology at Virginia Tech and a clinical
psychologist, is the Founder and Director of Recovery Efforts After Adult and Child
Trauma Clinic (R. E. A. A. C. T.). Specializing in clinical child and adult trauma,
his research and clinical efforts are focused on the assessment and treatment of
PTSD and related disorders resulting from traumatic events, with special interest in
the areas of recovery from disasters and cultural competence approaches to assessment and treatment of those suffering from a variety of traumatic events.

**Katharine Donlon** is a second-year doctoral candidate in the VT Clinical
Psychology Program with an emphasis on traumatic stress. Her research interests
concern the impact of the April 16 events on the Virginia Tech community as well
as on students' post-traumatic growth.

**Kelly Dugan Burns**, doctoral student in clinical psychology, is presently completing her internship as a pediatric psychology resident at Geisinger Medical Center in
Danville, Pennsylvania. Her clinical and research interests are in the area of childhood trauma and pediatric medical traumatic stress. Prior to her graduate training
at Virginia Tech, she coordinated clinical research for a project for families with
young children near Ground Zero in New York City following the September 11th,
2001, attacks.

**Kathryn Schwartz-Goel**, a VT psychology doctoral student, is interested in how
different types of loss following a trauma affect future functioning and what can
be done to promote positive adaptation following these events. She is currently on
internship at Park Place Behavioral Health in Orlando, Florida.

**Mary Kate Law**, a VT doctoral student whose research and clinical interests
were influenced by the 4-16 shootings, is currently on internship at the Veterans
Administration (VA) Maryland Health Care System, where she is involved in the
treatment of traumatic stress

**WHAT'S NEXT?**

Thus far, chapter authors have shared advice based on their experience with tragedies of scale. In the following chapter, school safety expert Michael Dorn, himself a victim of chronic childhood bullying, provides a look inside the personal world of individual victimization, how it affects learning, what schools can do to create safe learning environments, and how they can assist students in recovery from abuse and bullying.

# CHAPTER 11

# BURNED INTO MEMORY: REMNANTS OF PERSONAL VICTIMIZATION

MICHAEL DORN

Jacob's words seared like a branding iron. "I know what bullying is—that's when kids make fun of me because I am dying, can't run like they can, or when they tease me because I can't throw a ball like they do."

THE THOUGHT OF ANYONE SAYING SUCH HURTFUL THINGS TO another human being rattles me, and it would be easy for emotions to take over from reason. Over the course of my life, I have been shot at, cut with a razor, and, like most adults, have had my share of painful life experiences. However, Jacob's words hit me like a sledgehammer in the gut. If you knew Jacob, they would likely hit you just as hard. A truly impressive, successful, and compassionate child, he is indeed dying.

Jacob has muscular dystrophy and, at the age of seven, seemed to fully comprehend the lethality of his condition. He almost always wore a natural smile, exhibiting care and gentle concern for animals, and true compassion for other people. While Jacob had every reason to make excuses and to become an unhappy person, he conveyed a sense that he did not have time to waste on that sort of thing. He, instead, chose to make the best of every precious minute. We can learn a great deal from Jacob and from kids like him the world over.

## UNDERSTANDING THE ISSUE

Studies estimate that of the millions of K-12 students in U.S. schools, each year approximately one-third will experience some form of bullying and victimization (White House Conference on Bully Prevention, 2011). Some of these traumatizing incidents may be a series of acts of lower level harassment; others may be ongoing and intense. They range from verbal abuse (such as subtle put-downs and name-calling) to exclusion, physical intimidation, and violent assault. Perpetrators are adept at finding ways to try to make others feel powerless, humiliated, and unsure.

Media reports often portray bullying as a new and emergent phenomenon, yet it is certainly not new. The trauma experienced by those who are bullied received increased national attention during the 2009–2010 school year, in part because of the intensive media coverage of a number of suicides by students who had been subjected to ongoing abuse at the hands of their classmates. While the link between being bullied and the onset of depression and suicidal thoughts has been observed for decades, most Americans have not paid much attention to the problem. Reports of the suicide of Phoebe Prince, a pretty, 15-year-old high school freshman, however, put the crushing effect of this form of personal victimization into human terms. Phoebe, who had recently moved with her family to Massachusetts from Ireland, had been the target of relentless cyberbullying, harassment, and taunting by older girls in her school who resented her for dating a member of the football team. She reportedly asked her school administrators for help just one week before she took her own life (Kern, 2010), yet found herself unable to endure the torment any longer. The media devoted considerable coverage to this particular case after largely ignoring others like it over the years. Subsequently, other high-profile student suicides attributed to bullying have received considerable media coverage. Though much of this reporting is sensationalized and in some cases inaccurate, as is often the case with emotionally laden issues, the attention has generated heightened demands to combat an age-old problem. It has also increased understanding that victims of abuse need support from their parents, classmates, school, and community.

Fortunately, dedicated individuals have long been researching the problem (e.g., Olweus, 1978, 1979). Much has been learned about how to help prevent the abuse and how to aid its victims. Increasing public awareness of the traumas inflicted by bullies is an important step toward solving the problem. However, in some cases, popular talk shows have broadcast exposés that may have done as much harm as good. Several programs, for example, covered the suicide of a high school student who reportedly killed himself because he had been ridiculed due to his sexual preference. In reality, there was no indication that the young man was gay or bisexual. The coverage of the incident shifted the focus from his abuse by bullies and the long-standing mental health problems he had experienced to a question of sexual preference. This misdirection caused severe emotional distress for his mother, for school officials, and for others who were deeply pained by the young man's suicide.

A number of proven, inexpensive approaches to reducing abusive behavior have been available for quite some time. However, many of these are overlooked as people seek out "experts" from the growing field of school safety, a multibillion dollar industry that offers products and services ranging from security cameras, access control systems, training programs, and a host of other wares. These strategies, while well-intentioned, may at times be no more effective than the lower-cost approaches, and some can even be counterproductive. Books on the topic abound, and, while some are excellent publications, others promote content that is quite appalling.

With the variety of programs, technologies, and packaged approaches designed to help schools solve the problems of violence and victimization, educators and administrators must exercise caution before selecting an approach for their school. Each situation and each student body is different and requires careful analysis before any systemwide measure is adopted. It is as important to reduce aggressive behaviors as it is to support the victims in recovering the sense of basic safety that is prerequisite for learning.

At first glance, some innovative approaches to resolving abusive behavior appear quite promising, yet when implemented they may produce unanticipated harm. I recall attending what I thought to be an exceptional training program as a young police officer. The program involved a unique approach in which officers were trained to use marriage counseling techniques when they responded to all too common domestic dispute calls. Officers who conducted the program clearly believed in its merits and worked hard to convince us that we should take the time to help couples reconcile their differences when we responded to this type of call. However, I later discovered that not only was this approach ineffective, but that it had actually *increased* the chances that one of the domestic partners would subsequently be murdered. Though the officers who developed, implemented, and presented the program had the best of intentions, it turned out that their efforts were doing much greater harm than good. The situation illustrates the need to evaluate the effects of any program that is adopted.

Similar ill-advised efforts have been seen in some campaigns against bullying. Caring and dedicated adults who did not fully understand the dynamics of abuse have attempted seemingly promising approaches, in some cases even attempting to convert abusers into protectors of victims. Some advocate peer mediation, even though this approach, which has been shown to be quite effective in resolving some types of student disagreements (e.g., Cantrell, Parks-Savage, & Rhefus, 2007), is less likely to be successful in cases of bully-victim conflict. The dynamics of bullying are not merely a matter of a dispute to be resolved, but reflect a perceived imbalance of power between aggressor and victim. Indeed, by definition, *peer* mediation is a situation in which "students resolve conflicts through the facilitation of an equal (or peer)" rather than through the intervention of a parent, teacher, or administrator (Nix & Hale, 2007, p. 327). In cases involving bullies and their victims, a perceived power differential has already been established, thus finding a *peer* who can achieve resolution is unlikely.

Educators who understand the traumatic impact of this highly personal form of victimization realize that its resolution is an issue of moral and legal obligation and that its abatement can actually help improve academic achievement. Research has shown that bullying and harassment leave victims with reduced academic performance, feelings of being unsafe at school, symptoms of depression, and even suicidal thoughts (Bauman, 2008). In 2009, researchers in Britain reported a direct statistical correlation between bullying and lower levels of academic achievement (National Centre for School Research, 2009). In a review of research into school violence, Eisenbraun (2007) reported that victims of violence suffered a decline academically and socially and that the dynamics set in place by being bullied made it more likely that the victim would be shunned by peers and become a target of future violence (e.g., Beale, 2001; Gilmartin, 1987; Bulach, Fulbright, & Williams, 2003; Osofsky & Osofsky, 2001). Thus, the overall climate of the school itself can become increasingly toxic, as victimization increases the likelihood of further aggression as well as deteriorating academic achievement (e.g., Brockenbrough, Cornell, & Loper, 2002).

There seems little doubt: Victims are at risk of potentially devastating outcomes. What may be less clear is how educators can help end the abuse, and how they can assist students in surmounting the challenges they face in the experience. I'd like to offer a brief historical perspective to provide a context for assessing the dynamics of personal persecution and victimization and for developing strategies to help its victims.

## CRUELTY AS A HUMAN TRAIT

In the late 1930s, Nazi Germany began a systematic process of exterminating all special needs children. This mass murder of Germans by Germans to cleanse the Aryan race was a precursor to some of the cruelest and most systematic forms of mass murder ever carried out. Estimates are that as many as 13 million Jews, Gypsies, homosexuals, Poles, and others who were in some way viewed as a threat to the German race were shot, starved, worked to death, or otherwise brutally murdered (e.g., Childers, 2001; Evans, 2009).

While most Americans have a general awareness of the Holocaust, they are typically not as well versed on the many other instances of genocide perpetrated around the world over the centuries. Consider some of the better known examples: More than 70 million Chinese were killed during the reign of Chairman Mao, who worked tirelessly to ensure that every person in China witnessed at least one public execution as a means to instill the terror he needed to cement his power (Chang, 2006). The reign of Soviet Dictator Joseph Stalin saw the annihilation of between 20 to 60 million people (Klein, 2004; Fears, 2007). Describing statistically smaller, yet equally barbaric atrocities that have been carried out over the centuries in the Middle East, Asia, Europe, Africa, the Pacific Rim, North America, the Mediterranean, and elsewhere would take far more text than this chapter allows. But the point is, a review of ancient and modern history reveals that mankind has historically demonstrated that, in spite of the many good things people do to help others, ours can be a truly violent and cruel species (Diamond, 1999; Ali, Rodrigues, & Moodie, 1999).

When we consider how many people around the world do not respond to current situations of mass suffering in Darfur, for example, and how many still deny the Holocaust as a historic fact, it should not surprise us when an adult who should know better dismisses bullying with an offhand comment like, "Boys will be boys." Any competent historian can quickly provide dozens of examples of terrible instances in history where large numbers of people supported, endorsed, stood silently by, and otherwise failed to impede the efforts of one group of people to suppress and even exterminate those who were seen as different from them.

Of course, school bullying even in its worst form does not begin to rival the horrors of systemic, state-run campaigns of persecution and annihilation. However, it does exhibit some of the same group dynamics, namely a core lack of empathy and the tendency for people to stand by while others are harmed, most likely because they fear that they too will be victimized.

A counterpoint to this dark side of humanity is the courage and compassion of the many brave individuals who, at great personal risk, come to the aid of the oppressed. Before and during World War II, many people in Germany, France, Italy, the Netherlands, and other countries risked their own life in an effort to save others from the Nazi state. So too, today many individuals stand in support of victims and expose themselves to potentially harmful consequences at the hands of the abuser. Though as adults we realize that the risks taken in the face of state-orchestrated violence are infinitely greater, we should keep in mind that it takes considerable courage for an individual child to stand up for a classmate who is being tormented (Evans, 2009).

## LEVELS OF VICTIMIZATION

As with the many examples of human abusers across time and around the world, it should be clear that bullies—those who exploit, intimidate, exclude, and methodically subjugate others—are not unique to American schools nor are they limited to K-12 settings. Called "victimizing" in South Africa and without even a standard term to describe it in most Vietnamese schools, bullying has been identified as a problem worldwide. While media reports regarding student-on-student victimization might lead one to assume that schools in the U.S. are more violent than elsewhere, such comparisons are complex and often invalid. Americans dare to educate those who are not even offered education in some other parts of the world. In some countries, for example, students with special needs are routinely excluded from traditional schools. In many locales, student discipline is still the parent's responsibility, and students who exhibit what we would consider to be relatively minor disciplinary infractions are permanently expelled. Students in Vietnamese schools are graded for their behavior, as well as for their academic performance. Failure to conduct oneself properly can result in a failure to advance and, in extreme cases, can be grounds for expulsion from the school system altogether. It is difficult to contrast school safety in American schools with that in countries where a student who is caught with a gun or drugs faces torture, lengthy and barbaric prison conditions, or, in some instances, execution.

Parental involvement in their children's education varies in different parts of the world. Whether stemming from greater parent participation, more demanding standards of achievement, criminal justice sanctions, a greater appreciation for the opportunities that education affords, or disparity in access to an education, stark differences in the way educational services are delivered globally make it a challenge to compare international statistics related to matters of school safety.

Difficulties in understanding relative levels of lethal school violence in different settings also arise because of the manner in which incidents are reported. For example, if a child is assaulted and killed as he leaves the gates of his school in the United Kingdom, it will typically not be counted as a school-associated death as it would be in the United States. While total perimeter fencing of school grounds is highly unusual in the United States, it is the norm in many other parts of the world, including the United Kingdom, Israel, South Africa, and Vietnam. Simply relocating the assault by a few feet changes the data completely. Since it is not possible to accurately calculate the relative levels of lethal school violence, it is impossible to compare bullying and other forms of abuse in schools without standardizing what is meant by the term and the methods of measuring it.

My point in this discussion is not to minimize the significance of the problem in the United States, but to emphasize that understanding the implications of this global concern is difficult. Merely identifying countries or school districts that appear to have lower incidence of bullying and then adopting their approach cannot be assumed to produce the same result elsewhere. A deeper understanding of the issues, educational climate, cultural norms, demographics, legal system, and even a simple definition of terms is a prerequisite.

## A CULTURAL CHANGE

In general, we in the United States long for a simple solution or quick fix for our problems. We would like to purchase a product, take a pill, implement a program, or in some other fashion work out our troubles quickly and easily. But solving the complex societal problems we face—whether it is suicide, substance abuse, gangs, bullying, or any of the myriad of other perplexing dilemmas—requires complex thinking and comprehensive approaches. This challenge is true with regard to all forms of violence prevention, intervention, and recovery.

We are not alone in this desire for simple solutions. One South African principal, for example, told me that he had addressed the problem in his school by appointing bullies to mediate bullying situations, an approach that is counterproductive at best, and dangerous at the worst.

Bullying is a pervasive social problem requiring a coordinated approach to affect significant positive change. Hosting student assemblies or putting up posters in the halls alone will not achieve lasting improvement, and students who are bullied will find no solace in such simplistic measures. An effective approach to any school problem involves a number of key elements, including efforts to educate the entire school community. To successfully address a deeply rooted societal problem, a cultural change is required. This type of change involves consistent efforts to educate staff, students, parents, and the community about the dynamics and the consequences of the problem. While it is not always easy to change the way people think about societal issues, efforts to educate them about bullying can help reduce its occurrence. In addition, changing the blame-the-victim mentality is as crucial to prevention as it is to recovery.

Many people remain locked in their own beliefs about the situation because they view it through the lens of their own experience. Encouraging the breakdown of barriers to perception is an important first step in bringing about significant cultural change. As an example, consider the following case. A part-time, school safety employee in Texas made it a practice to speak to every student who had not been picked up by the end of her shift to verify that they had transportation home. On one occasion, she found a student who was sobbing, having a complete breakdown. She summoned the principal and two school resource officers, and after about 10 minutes, the girl calmed down enough to explain that she had been bullied at the school every day for the previous two years and that she simply could not take it any longer. She related that she could not return to the school ever again. The principal was astounded that the student could have been bullied in his school for so long without his being aware of the situation. When he asked her why she had not reported the abuse, she told him that she had told her mother about it when the bullying had begun, and that her mother had told her to learn to be tough. The girl's mother was a teacher in another school in the district. This well-meaning but ineffective and simplistic advice is a common response by students, school staff, and parents. Fortunately, a dedicated and alert school employee had noticed the indications of a problem and followed through properly. Commonly, we rely on our own experiences, as this parent did, and the advice we give may not be appropriate to the student or the situation.

Efforts to improve school culture focus on educating the students about bullying; however, everyone who is associated with the school—teachers, administrators, custodial staff, bus drivers, parents, guardians, everyone— needs reliable information and guidance. It is astounding to learn that a veteran school employee is largely ignorant about the problem, but what to do about bullying is not a topic that is routinely covered in teacher/principal preparation courses or recertification programs. A school administrator in Oregon, for example, was stunned at the manner in which a building principal in her district handled a case where her daughter had been traumatized by bullies. When she told her colleague that her daughter had been savagely, verbally abused by several girls at the school, the principal said that she would resolve the situation immediately. The daughter later called her mother in tears demanding to be picked up from the school. She informed her mother that the principal had placed her and the girls who had been bullying her in a room with instructions not to leave until they had worked things out. The girls had cursed her and vowed revenge against her for reporting them. The girl told her mother she would never return to that school and wanted to enroll in a home school program. Her mother was not only astounded that a fellow administrator did not understand how poor a response this was, but was also shocked that the principal defended her approach as the best way to deal with bullying.

Bringing about positive cultural change takes time and continued commitment. Student assemblies, for example, are limited in what they can accomplish, and such activities should be seen as only one part of a more comprehensive approach. Similarly, a program that is developed or implemented through a short-term grant is unlikely to produce lasting improvements unless the innovations generated through that effort continue after the grant has ended. To maintain the momentum of change, efforts to educate parents, students, and staff must be systemic, sustained, and subject to ongoing evaluation. An approach may work for a while, but as the situation changes, it needs to be revisited.

While press coverage of student suicides has helped many people develop greater empathy and concern, it is vital that the information they are given is accurate, culturally relevant, and up-to-date. School officials should carefully vet any information that they distribute. Repeating common myths and misperceptions can do considerable harm, and although the Internet makes it easy to find information, it must be carefully scrutinized for accuracy and intent. [A few of the more highly regarded sources of information are listed in Chapter 14.]

## ASSESSING SCHOOL SECURITY, CLIMATE, AND CULTURE

Students need to feel safe if they are to learn, and any approach to school safety and violence prevention needs to be continually monitored and evaluated. Surveys of students, staff, and parents that include questions about victimization are requisite to gauging the level of incidence as well as the progress that has been made with prevention measures. Educators will be able to identify the nature and the extent of problems in their schools only by asking relevant questions in ways that are clearly understood. It is fairly common for surveys of students and staff in the same school to show a dramatic difference in the perceptions of each group when it comes to the question of school safety.

One relatively simple approach to discovering the extent of the problem is for school officials to track instances of bullying by time of day and location. For example, the Clark County School District in Las Vegas, Nevada, achieved significant improvement in a pilot project that asked students to indicate hotspots of inappropriate student behaviors on a web-based map of their school. Administrators could then view color-coded graphic displays of their building to see when and where problems were occurring. By assigning specific staff members to supervise in the areas and times of greatest concern, pilot schools were able to reduce problematic behaviors by approximately 50 percent. In contrast, I know of no situation in which simply installing school security technology alone reduced incidents by this amount.

In my work with schools around the globe, I have seen improvements in student supervision combined with prompt and appropriate intervention of adults is one of the most powerful, inexpensive, and easy-to-achieve approaches to school safety. Security cameras can be a useful investigative tool and offer some deterrent value, but there are many cases in which school shootings and other major acts of violence have been recorded, but not prevented, by security cameras. I do not mean to imply that security cameras are ineffective, but rather that they must be utilized appropriately and effectively. Cameras, like other types of security technology, should always be used in concert with proven methods and timely educator response. Properly conducted assessments combined with meaningful follow-up activities, such as increased student supervision and models for improved communication, can help to reduce the frequency and severity of victimization and increase the effectiveness of intervention and recovery strategies.

In order for students to be able to learn and to make the most of their school years, they must first feel at ease in their school environment. A student who fears for his or her personal safety or one who is suffering from the trauma of chronic abuse and intimidation will not be able to benefit from the instruction that a school offers. A holistic approach to creating safe and secure schools can reduce the level and scope of traumatizing experiences and dramatically improve school climate, culture, and academic performance. Indeed, many K-12 school systems and college campuses are utilizing advances in safety and security as school improvement tools.

## UNDERSTANDING HOW VICTIMS PERCEIVE THEIR EXPERIENCE

Three factors determine how victims interpret and understand the experience of being abused or threatened. These factors—frequency, severity, and the perceived level of shame—will influence how an individual responds to victimization. For example, a student who is repeatedly called derogatory names for a period of several months may become severely depressed and withdrawn. Another student might be just as severely impacted by only a single instance of more humiliating behavior, such as being physically assaulted in front of peers. In many instances, victims of bullying face both high frequency and high severity of aggressive behaviors. A student who has witnessed or survived a traumatic experience in the past is more sensitized to victimization in the future, and when a high degree of perceived shame is added, the impact on a young person can be even more devastating.

Adults often forget the depths of the peer pressure they themselves experienced as a student. To many children and youth, normal peer pressure can become a powerfully degrading factor when viewed in the context of humiliations they have experienced in front of others, whether in person, online, or both. Clearly, the problem of cyberbullying, which records and disseminates a student's humiliation to countless thousands via the Internet, has intensified the potential for traumatic victimization. The Internet also allows a vast number of young people to bully single victims with relative ease and a feeling of immunity from consequences for harassing distant victims. As we have seen, the perceived level of shame has proven to be extremely painful to victims tormented via the technologies of today.

In addition to the distress caused by the bully, additional torment is caused when bystanders fail to act, friends turn away, parents minimize the experience, or educators respond inappropriately. Loss of face before classmates combined with loss of trust in adults can further isolate and plague a bullied child. It should be noted that witnesses to another's abuse often experience anxiety and a level of trauma as well.

### HOW TO HELP

Creating a school culture that does not tolerate abusive behavior helps establish a positive environment for teaching and learning, yet ending the bullying is only part of the equation. Helping victims recover and rebuild their sense of safety and confidence deserves equal attention. Solely attending to the perpetrator does not take away the negative effects of having been abused.

At the 2011 White House Conference on Bullying Prevention, educators, parents, researchers, and innovators from the public and private sectors focused their attention on stopping the abuse and on supporting the abused. Participants advocated that administrators develop clear policies and simplify procedures in order to streamline reporting and facilitate prompt sharing of information so that when a student reports a problem, it will be attended to immediately. If students are encouraged to report incidents of abuse, they must see that their actions produce results, that adults are allies who will work with them to improve the situation, and that they will be safe from retaliation.

While bullying has been around for a long, long time, current iterations are perhaps more insidious and complex. Cyberbullying has the potential to follow the student everywhere, via cell phone, social networks like Facebook, texting, e-mails, and so forth. Students experiencing this type of abuse may be reluctant to report the problem to parents or teachers for fear that they will lose access to their computer or their phone. While simply "pulling the plug" on the media may seem a quick solution to the problem, in effect this response punishes the victims by disconnecting them from social contact with their friends and the support they need. Instead, responsible adults need to assure the student that they will not lose connectivity, but that the situation will immediately be reported to those who can appropriately and safely intervene. Social network sites are becoming more responsive to the problem of cyberbullying, and many are instituting increased safety features to allow for reporting of abusive posts, with the potential for those who continue to harass others to lose access to the network.

Parents need to know how to monitor websites where videos are posted, social network sites, and any other media where abuse may occur. All adults are responsible for protecting all students and need to be alert to signs of victimization. They also need to know how to take appropriate action, engaging the students themselves as well as other involved parents, in achieving satisfactory resolution.

While students can be harmed by the experience of being bullied and abused, a close and supportive environment can foster resilience and increase the likelihood they will be able to cope effectively. A study of elementary and secondary students revealed that support from teachers and parents helped buffer children who had been bullied from experiencing symptoms of depression (Conners-Burrow, Johnson, Whiteside-Mansell, McKelvey, & Gargus, 2009). A supportive family was among the most significant factors that influenced how well young people handled life's stressors; the study also concluded that support from teachers helped prevent symptoms of depression when the children were not getting the support they needed at home. Especially informative was the following:

> The indicators of social support measured in this study included simple things such as listening, offering advice and help in problem solving, and showing that they care. By engaging in these behaviors, parents and teachers may be able to help protect even the most at-risk children and adolescents from experiencing depressive symptoms. (p. 603)

Simple actions—listening and demonstrating care—do make a difference. Supportive friends are helpful, but informed and caring adults, whether at home or at school, can provide the essential assistance, guidance, and support that only years of life experience can create. Parents as well as teachers thus need a solid appreciation for the dynamics of bullying in order to be more effective in helping victims to problem-solve these extremely complex and painful life situations.

Students who experience severe and/or chronic bullying may need more than the loving support that parents, teachers, and friends can provide. Some children will feel deep emotional and psychological wounds and require counseling to help them recover. Unfortunately, many students neither seek nor receive this type of assistance. On many occasions, after speaking at a safety conference about my own experiences as a victim of bullying, I have been approached by adults who tell me their own stories, in tears, of how badly they were bullied 10, 20, 30, or more years before. Many of these adults tell me they have not recovered from the pain and humiliation they endured as children, but report that they have never sought assistance from a mental health professional to help them deal with their trauma. If even adults do not seek mental health assistance when their lives have been so severely affected, how can we expect young people to ask for help on their own? Most do not know how to access mental health resources, or they refrain from seeking help because of the social stigma attached to needing therapy. While many school districts lack adequate mental health resources, educators can build strong relationships with the mental health community, to clarify the resources that are available and to learn when it is advisable to initiate referral so that victims can get help.

Some adults who have dealt successfully with severe bullying during their own childhood are willing to talk freely about their experience and advocate for others who are suffering. Often, they report that they can clearly recall the actions of friends and classmates who were willing to stand up for them when they were being tormented. They also will tell of adults who provided memorable support that helped them persevere. Just as the painful words and deeds of aggressors can remain locked in memories for many decades, so too can the kindness and bravery of others. Consider the following:

When one enters Yad Vashem, Israel's official memorial to the Jewish victims of the Holocaust, certain sights are immediately burned in memory. A railway boxcar emblazoned with swastikas that projects from the hilltop is among them. The railcar was a gift to the new state of Israel from the Soviet Union, along with the meticulous records of the Jewish men, women, and children who were transported to their deaths, sometimes after paying for a ticket to take them to the safety of a neutral country. Many visitors to Yad Vashem are equally touched by the avenue of trees that surrounds the exterior grounds. In dedicating this memorial to those who perished in the Holocaust, Israelis have also worked tirelessly to identify the thousands of Christians, Muslims, and people of other faiths who risked their lives to rescue Jews from certain extermination by the Nazi regime. Each tree along the avenue serves as a physical recognition of the "Righteous among Nations," those who courageously stood against evil even at great personal risk. The names of those selfless and brave people who protected others helped to prove that not all will stand by and silently assent to persecution. This tribute to those who did not ignore the suffering of the Jews serves as a reminder that each of us is responsible for our own actions and that "if one was to build a future in a world where Auschwitz had become a real possibility, it was essential to emphasize that Man was also capable of defending and maintaining human values" (Yad Vashem, 2011).

Whether in the face of state-run genocide or the case of a school-yard bully, the heartless behaviors of people who abuse and exploit can be seen in stark contrast to the decency of those who refuse to stand by and allow others to be tormented. Though the levels of violence and the bravery required to counter it pale in comparison, the significance of bystander behavior bears similarities that we should remember if we hope to protect people from those among us who seek to intimidate and destroy without conscience.

## JACOB'S GRACE

While the problem of bullying is likely to remain, there is hope for the future. In my travels, I meet many educators, parents, school resource officers, mental health professionals, elected officials, government workers, and members of the general public who are committed to changing the situation. I have also met thousands of young people who inspire me with their efforts to make a difference in the lives of others.

Though it might seem easy to focus on negative accounts of bullying, there are many indicators of progress if we look for them. Improvements in the quality of anti-bullying efforts, increased media attention, appropriate legal action, and better communication between adults and students are among the factors that are bringing positive change to many schools. Likewise, a rising awareness of the traumatic impact of bullying on victims is increasing the commitment to help them transcend their abuse.

Jacob, the young person I introduced you to at the beginning of this chapter, handled his bullying experiences with a remarkable degree of maturity and grace. When I urged him to tell his teachers and principal what he was experiencing, he asked me why he should do that. From my own history, I knew that many kids don't report bullying because they don't feel the adults can or will handle the problem well. I pointed out to Jacob that the adults in his life could help him, and he told me that he knew that they could. He went on to explain how the principal and teachers did not allow bullying at his school, and he related how they taught the children why it was okay for kids to be different. He told me that the teachers at his school watched the kids closely and that he felt safe there. He then enlightened me that the abuse he had experienced had not happened at his school but rather in other settings. He told me that he could handle the terrible words of other kids because he knew that his principal, his teachers, and his mom and dad loved him. Jacob reminds us that context is key to understanding.

Bullying can sear painful memories into its victim. At the same time, the acts of valor and even subtle kindness are gifts to be remembered and cherished as the decades pass. Both the destructive and the gracious acts of others can be etched into the memories of those who experience them. Like countless others who have been on the receiving end of severe physical and emotional abuse as well as benevolent acts of grace that can help one's soul survive, I have many memories that do not seem to fade much with time. The perspective through which I view these experiences is ever evolving, and I cannot forget having my head smashed into a hard elementary classroom floor any more than I can forget the valor of my teacher who stood up to her building principal on my behalf so I could get help for my learning disability. While I recall the pain and humiliation at the hands of others, it is equally as impossible for me to forget the fact that one wonderful teacher advocated for special testing for me that was not covered in the school's budget so that my learning problem could be identified. When it was determined that I had dyslexia, she worked to find a scholarship for me so I could attend a treatment program. Because of her support, I was able to develop skills and strengths that allowed me to overcome the challenges I faced. They said that I would never learn to read. Yet, because of this one teacher's perseverance and kindness, I am now not only able to read books, but also to write them. Throughout history, such grace has always been in demand.

Different kids respond differently to the experience of abuse, and the support that adults provide is often the determining factor in its overall effect. Jacob shows us that adults have it in their power to create environments where bullying is rare and that, should abuse happen in spite of their efforts, the victims will be supported and nurtured as they work their way through. We can all make excuses about a lot of things in life, but Jacob reminds us of our role in protecting kids and that each of us can make a difference.

## REFERENCES

Ali, J., Rodrigues, L., & Moodie, M. (1999). *Jane's chemical-biological defense guidebook*. London: Jane's Information Group.

Bauman, S. (2008). Victimization by bullying and harassment in high school: Findings from the 2005 youth risk behavior survey in a southwestern state. *Journal of School Violence, 7*(3), 86–104.

Beale, A. V. (2001). BullyBusters: Using drama to empower students to take a stand against bullying behavior. *Professional School Counseling 4*, 300–305.

Brockenbrough, K. K., Cornell, D. G., & Loper, A. S. (2002). Aggressive attitudes among victims of violence at school. *Education and Treatment of Children, 25*, 273–287.

Bulach, C., Fulbright, J. P., & Williams, R. (2003). Bullying behavior: What is the potential for violence at your school? *Journal of Instructional Psychology, 30*, 156–164.

Cantrell, R., Parks-Savage, A., & Rhefus, M. (2007). Reducing levels of elementary school violence with peer mediation. *Professional School Counseling, 10*(5), 475–481.

Chang, J. (2006). *Mao: The unknown story*. New York: Anchor Books.

Childers, T. (2001). *A history of Hitler's empire* (2nd ed.). Chantilly, VA: The Teaching Company.

Conners-Burrow, N. A., Johnson, D. L., Whiteside-Mansell, L., McKelvey, L., & Gargus, R. A. (2009). Adults matter: Protecting children from the negative impacts of bullying. *Psychology in the Schools, 46*(7), 593–604. Retrieved March 13, 2011, from www.interscience.wiley.com.

Diamond, J. (1999). *Guns, germs, and steel.* New York: W.W. Norton & Company.

Eisenbraun, K. D (2007). Violence in schools: Prevalence, prediction, and prevention. *Aggression and Violent Behavior, 12*(4), 459–469.

Evans, R. J. (2009). *The Third Reich at war.* New York: Penguin Press.

Fears, J. R. (2007). *The wisdom of history: The great courses.* Chantilly, VA: The Teaching Company.

Gilmartin, B. G. (1987). Peer group antecedents of severe love-shyness in males. *Journal of Personality, 55,* 467–489.

Kern, T. (Summer 2010). The anti-bully campaign: Therapists share their tactics. *Annals of the American Psychotherapy Association.* Retrieved March 7, 2011, from www.americanpsychotherapy.com.

Klein, S. (2004). *The most evil dictators in history.* New York: Barnes & Noble.

National Centre for Social Research. (2009). *The characteristics of bullying victims in schools.* Retrieved March 13, 2011, from http://www.education.gov.uk/publications/eOrderingDownload/DCSF-RBX-09-14.pdf.

Nix, C. L., & Hale, C. (2007). Conflict within the structure of peer mediation: An examination of controlled confrontations in an at-risk school. *Conflict Resolution Quarterly, 24*(3), 327–348.

Olweus, D. (1978). *Aggression in the schools: Bullies and whipping boys.* Washington, DC: Hemisphere (Wiley).

Olweus, D. (1979). Stability of aggressive reaction patterns in males: A review. *Psychological Bulletin, 86,* 852–875.

Osofsky H. J., & Osofsky, J. D. (2001). Violent and aggressive behaviors in youth: A mental health and prevention perspective. *Psychiatry, Interpersonal and Biological Processes, 64,* 285–295.

White House Conference on Bullying Prevention. (2011). Retrieved March 10, 2011, from http://www.whitehouse.gov/.

Yad Vashem. (2011). Holocaust Martyrs' and Heroes' Remembrance Authority. Retrieved March 7, 2011, from http://www1.yadvashem.org/.

\*   \*   \*

## CONTRIBUTOR NOTES

**Michael Dorn** is Executive Director of Safe Havens International, a global non-profit school safety center dedicated to making students safer anywhere in the world. During his 20-year public safety career he served with the Mercer University Police Department and was Chief of the Bibb County, Georgia Public Schools

Police Department, School Safety Specialist for the Governor's Office of Georgia, and Lead Program Manager for the Terrorism Division of the Georgia Office of Homeland Security. A graduate of the 181st session of the FBI National Academy, Michael is a passionate and tireless advocate for children and youth and has authored over two dozen books on school safety. Driven by his childhood experiences as a victim of severe bullying, Michael's work has taken him to Mexico, Central America, Canada, Europe, South Africa, Asia and the Middle East. Michael can be reached at www.safehavensinternational.org.

**WHAT'S NEXT?**

In the following chapter you will have an opportunity to learn from faculty, students, counselors, parents, clergy, and others who have advice to share based on their own experience with trauma and recovery. In addition to recommendations for what helps in the aftermath, they reveal what they found hindered or delayed recovery within their school community.

CHAPTER 12

# A ROAD TO NORMAL

## CAROLYN LUNSFORD MEARS

> There was an underlying pride that no matter what was said, this staff
> quietly and with little fanfare or public face had saved over 2,000 lives.
>
> Monette Park
> Social Worker, Columbine High School

A RECURRENT THEME IN THIS BOOK HAS BEEN THAT traumatic experience
affects functioning of individuals and institutions and as a result, reclaiming school in the aftermath requires reasoned adjustment to meet changed
needs. Ever since the Columbine tragedy, I have worked to identify what
helps people recover from community-wide trauma. This chapter is based
on almost 150 hours of interviews with educators, counselors, students, parents, clergy, law enforcement, and victim advocates. I am humbled by their
willingness to reflect on troubling events for the purpose of helping others,
and I want to thank them here for their assistance. To protect their identity,
I refer to them in this chapter by their role rather than by name, unless they
have expressly given me permission to quote them directly or have been
identified by name in other publications.

While trauma is, by definition, disturbing and life-changing, for many
of its victims, the outcomes can be transformative rather than debilitating.
One of the lessons from trauma that I especially want to emphasize is that
it is a mistake to stereotype survivors as forever broken by their experience.
Indeed, while traumatic memory is never erased, many transform their loss
into personal growth and a commitment to living a life of purpose.

> [Survivors] possess a special sort of wisdom, aware of the greatest threats and
> the deepest gifts of human existence. Life is simultaneously terrifying and
> wonderful. Their traumatic experience was undeniably agonizing, and yet,

having successfully struggled to rebuild their inner world, survivors emerge profoundly and gratefully aware of the extraordinary value of life in the face of the ever-present possibility of loss. (Janoff-Bulman, 1999, p. 320)

Those interviewed for this book generously shared their experiences and their suggestions for schools and communities. Some of their advice has been covered in detail in other chapters, and to avoid unwarranted duplication, I have chosen to focus this chapter on additional or alternative insights. In doing so, I want to acknowledge the power of all the stories that were shared with me, even if I have not included them here. These points of common ground clearly validate the significance of trauma as a concern for educators everywhere.

While the discussion in this chapter is framed in terms of large-scale events, many of the insights have relevance for individual cases as well. A community-wide disaster often attracts an outpouring of support and resources, but the needs of individual students who experience loss or victimization should not be overlooked. The scale is different, the scope of disruption more limited, but some of the recommendations for reclaiming school can be applied to the challenge of helping all students resume their educational path.

## LEADING IN TIMES OF TRAUMA

Adept leaders recognize and prepare for a wide range of possibilities, including the potential for crisis. Anticipating and planning for the special needs that result from trauma increase the likelihood for positive outcomes of even the most devastating experiences. Indeed, leadership before, during, and after a crisis often determines the rate of recovery. Responding to a crisis requires the same critical strengths of leadership that are fundamental *before*; and long-term recovery in the aftermath depends on those strengths being sustained over an unpredictable and often extended period of time. In many ways, the aftermath can be more difficult than the initial crisis, for the resulting anxiety and uncertainty are often compounded by secondary, disruptive events.

Chances for recovery are enhanced by coordination of efforts with others, including people and institutions outside the usual realm of daily educator practice—law enforcement, victim services, crisis responders, and medical and psychological support providers in the larger community. Among the first responses to a disaster will be the activation of emergency protocols and communication networks. Trusting relationships, clear communication, and working agreements built in advance of need increase the quality of the response and can minimize damage and losses.

Crisis communication technology and procedures for responding to a variety of incidents should be designed, tested, and practiced. A comprehensive plan for notifying and reuniting families with loved ones is essential, and details of emergency procedures should be included in information packets that go home at the start of every school year. They should be discussed at staff meetings and drills; they should be posted on the school website; and they should be reviewed and updated throughout the year. Substitute teachers, volunteers, and others who are regularly in the building or on campus also need the information.

Rehearsing emergency protocols in varied situations can prepare students and teachers to make on-the-spot, critical decisions. Evacuation procedures should be conducted under diverse conditions, since situations will change. A parking lot may be undergoing resurfacing, for example, or a building exit may be closed during remodeling. Students need to practice escaping from the building and then going to the designated pick-up spot once they are out. Assembling on the athletic field is not a sufficient strategy.

What actually transpires during an emergency will not look anything as orderly as what is written into the plan. Panic ensues among those exiting the building, while anxious family members descend on the pick-up point, snarling traffic, blocking roadways, and possibly hindering access of emergency vehicles.

Even the chaos of a false alarm can prove a valuable lesson. When an erroneous report of a person with a gun on a school campus caused a crisis situation in a small town, faculty, students, and family members experienced the turmoil of evacuation. Lessons from this event were shared and plans adjusted so that several weeks later when a real emergency occurred, the response was more orderly.

Plans for evacuating students with limited mobility or cognitive delays should be drawn up with input from parents or caregivers. If appropriate, a designated staff member and back-up should be assigned to aid the students in exiting the building. It is important to assign sufficient staff to this role, and to make sure that substitutes know of their responsibilities to assist during an evacuation.

Decisions made in times of crisis have ongoing implications for the entire school community. While advance planning to identify potential challenges of a disaster experience can help leaders prepare for many contingencies, a real disaster is merely the first of a series of crises, a string of critical incidents that doesn't end when the police arrive on scene or when the building is cleared and classes resume. Carrying their own personal trauma, educational leaders must oversee the usual business of school with the added demands of responding to heightened student and staff needs, intense scrutiny by media, police investigations, questions of liability, threats to the school, and the potential for loss due to changes in student behavior (e.g., aggressiveness, risk-taking, substance abuse).

Recovery comes in stages. After the immediate crisis is resolved, leaders must identify the losses and implement plans to access resources that are prerequisite to restoring function. Buildings damaged in the incident require thorough inspection for structural and operational safety. Allocation of space and equipment is complicated when a disaster makes multiple facilities unusable. Whenever school functions need to be transferred to sites outside the affected area, increased transportation costs, rescheduling of courses, and overcrowding of classrooms often result. Large-scale events create the potential for systemwide disruption, affecting not only instructional facilities but also administrative offices that handle such critical functions as acquisitions, payroll, student records, human resources, district management, and technology support. Strategic planning to address comprehensive crisis needs can better prepare a district or university to resume these functions more quickly.

## MEMORIALS, RITUAL, AND RELIGION

In the early days after a tragedy has engulfed a community, local leaders may be called on to plan or participate in memorials, candlelight vigils, and other remembrance events. Such ceremonies allow people to come together for mutual support and can foster a sense of togetherness as each tries individually to make meaning from the experience. The most beneficial of such occasions are those that support the families of the victims in their grief while promoting healing for the survivors and all who were touched by the tragedy. Achieving this delicate balance requires considerable forethought and sensitivity.

Leaders should be prepared for expressions of religious belief as part of the healing process and, at the same time, attempt to avoid divisive expressions that cause conflict. Accommodating people of different religions, as well as those who reject religion altogether, can help avoid long-lasting discord. Some ministers who came to Columbine from outside of the area participated in public ceremonies and brought solace to members of their faith. However, others who did not share the same religious tenets viewed some remarks with alarm. While finding comfort in the love of Christ may have been welcomed by Christians, for example, the message, which was intended to be inspirational, was hurtful to others. Noted Denver Rabbi, Steven Foster was quoted in an article about the controversy as saying, "The entire community was invited to come and mourn. And then it turned into an evangelical prayer service.... The issue was one of insensitivity to the kind of statements being made that were exclusively directed not just to Christians but to fundamentalist Christians" (Zoba, 2001, p. 55).

Expressions of core faith, ritual, and belief are, of course, appropriate to programs that are held at a church service or church-sponsored event. Yet some members of the local clergy observed that outsiders flocked to the community to promote their own theological agendas, and in a way "hijacked the public process by not respecting the different faith traditions that existed within the community." Some local ministers I talked with felt that for a public event or school-oriented memorial or program, the standard needs to be absolute inclusivity, regardless of religion, cultural tradition, or philosophical creed. Setting aside differences, if only for the moment, can help the school and entire community come together and express mutual support.

While faith frameworks can significantly contribute to the healing process, school leaders will be challenged to allow faith to be expressed in ways that are not considered offensive by those who have different beliefs. To avoid religious "fault lines" from developing, leaders are encouraged to work with district officials to strategize a policy that respects both individual freedoms and legal mandates governing separation of church and state. This type of discussion—in advance of need—will allow the development of plans with the kind of reasoned discernment that is difficult to achieve in the immediacy of trauma.

Policies need to be established regarding the presence of religious support within the school building after a tragedy. Some students interviewed for this chapter felt that religion sustained and supported them in their grief, and some told of being distressed when other students followed them down the hall in school, trying to convert them to their faith even after being asked to stop. Many school districts have policies against such actions on school grounds, while at least one official I spoke with described making a space available in a public school where religion could be freely discussed. Accommodating the disparity of parent and community responses will be the challenge, one that should be anticipated in the planning process.

Anniversaries of large-scale tragedies often honor and attest to the powerful loss that has been experienced. A counselor reminded me that marking the anniversaries of the tragedy or installing memorial plaques poses tough questions of how to acknowledge the loss without reopening wounds. Some may want to honor the anniversary with an emotional activity that recreates the intensity of caring that was demonstrated in the original experience. Others may avoid and decry such programs. Likewise, physical memorials or markers should be carefully planned with the best interests of all in mind. When a marker or structure is placed on school grounds, the students, faculty, staff, and families need to be involved in the planning.

## WHEN AND HOW TO RETURN?

Timing a return to classes is a delicate matter. While returning to a routine can be beneficial, it is difficult to know when people are ready. Among students I interviewed, some said the return to classes was much too soon, and others said they couldn't wait to get back. One student applauded his return: "I got a lot of comfort by being able to throw myself into my school work." Another felt returning to classes only a week after the tragedy was too soon: "I felt a lot of pressure to get back to normal but we were so traumatized by what we had heard and seen. I felt cornered and being pushed, as if I was holding other people back because I wasn't ready."

Teachers I interviewed welcomed the chance to be with their students again: "There was a lot of hugging going on. I just needed to be with the kids." Over the course of the following year, however, the stress of teaching in the highly energized environment began to take its toll and many longed for relief.

Developing contingency alternatives, if not plans, for holding classes in alternate locations can facilitate a return even after a building has been damaged. Knowing in advance what space and resources will be needed can expedite the process. Once on site, administrators, staff, and counselors need to assess the physical facilities and be alert to sensory stimuli that could retrigger the trauma. A Columbine librarian told me that when sessions were held at nearby Chatfield High School, students were reluctant to go into the library because of the violence that had been committed in the library at Columbine. She invited students in and showed them the location of the back exits in an attempt to reduce their anxiety. Elsewhere in the building, students were startled by a classroom door that banged every time it closed. A section of the roof "popped" loudly as it heated up during the day. These sudden noises brought back excruciating reminders of the sounds of gunfire. Teachers warned students about these sounds in hope of alleviating the effect, but the triggers were unnerving, nevertheless.

For survivors, any trauma in school or on campus will forever be associated with the physical location where it occurred and with all sensory stimuli that were experienced during the event. Large-scale tragedies may necessitate reconfiguring classrooms, redecorating hallways, and even repurposing entire buildings. Administrators and counselors tell of spending hours with mental health practitioners to plan new uses for rooms associated with a tragedy, selecting a "peaceful" color of paint for refurbishing the building, and choosing a new sound for the fire alarm. Before making significant decisions regarding the redesign of physical space, administrators are encouraged to give all affected individuals an opportunity to express an opinion about major changes. While it will not be possible to satisfy every point of view, at least people will know that they were invited to participate in the process.

Allowing family members to accompany students and school personnel on a tour of the building before classes resume can ease re-entry. One evening, with counselors available to lend support as needed, Columbine students, teachers, and their families were able to explore the building at their leisure, without being pressured as they would be during a regular school day. I remember walking through the cafeteria with my son, who paused and picked up a broken piece of floor tile that was left from the remodeling. He asked one of the aides if he could have it—it was part of the old floor, from a time prior to the tragedy. Its significance was a symbolic reminder that in restarting a school, it is important to honor what came before.

Parents shared with me how much they valued the chance to walk through the building with their sons and daughters since it gave them an opportunity to listen to their child tell what was, at that time, *tellable* about their experience. This occasion to tour the school unimpeded helped families better understand and benefited them as well. Following is one parent's description of the importance of this experience, written in the form of an excerpted narrative (See Appendix for an explanation of the development of narratives for gateway research [Mears, 2009]).

They got the school cleaned up,
And one night they opened it
For kids to take their parents in.
We went and
We saw where our child had been.
That was very important for us.
You have all of these questions as a parent,
As to what your child experienced.

So you want to understand it.
You have an overwhelming need to know.
You can't ever experience what they experienced,
But you had a need to understand it.
Somehow seeing the place helps you visualize
What it must have been like.

I wanted to go back in,
To see the place,
To look out at what he looked at
To sit where he sat,
To stand in his shoes for the three-hour period that he was in that room,

So I could imagine the fear.
That is what my child experienced and
I wanted to understand.

So that's what we did.
We went into the room,
Then we could understand,
To put the pieces of the puzzle together.
Someone had placed a rose in front of the door where he had been.
It just represented the pain.
It was a tribute to the pain.

What was important is that we went in with him,
And he pointed out,
This is where I was,
This is what I saw out that window,
The sharpshooters were over there,
But we didn't want to stand up.
We didn't know who was going to see us—maybe the shooters.
And this is what we could hear outside of the door.

It was so important that he share it with us as his parents.
Then we went to the other rooms to see
What other kids had experienced.
And of course we walked by the library.
It was sealed off, but it felt like a tomb.
It was like sacred ground.

                                              (Mears, 2009, pp. 7–8)

When students, faculty, and staff physically return to school and resume classes, they are beginning to psychologically reclaim their school, and the day of the return deserves special recognition. A triumphant return can help students begin to see themselves as victors instead of victims. Students at Columbine were welcomed back with an all-school rally in the parking lot, while hundreds of parents and family members formed a loving circle around them to show their support and, equally important, to protect them from the prying eyes of the media. As students entered the building that morning, they were greeted as returning heroes by their teachers and by the mother of a student who had been killed. Their bravery under extreme stress was acknowledged, and the message was clear: *You can do this and we are here for you.*

Some counselors advocate that administrators consider an option for a shortened school day for the first few days to lessen the stress of the return. A modified schedule can be accomplished by reducing the length of class periods or perhaps by holding morning classes one day and afternoon classes the next. As educators have pointed out, not much will happen academically in the first days back. At this particular time, the school's primary functions are to provide a structure that at least *resembles* a routine and to serve as a place for students and school personnel to interact. In the aftermath of trauma, flexibility becomes the standard. Some exposed to school violence will never be able to return to the classroom. Faculty and staff may need options for alternate assignments, early retirement, or additional personal leave to allow time for recovery. Students unable to return will require alternate educational services. Among the options that have proven successful in a variety of situations are online courses, home-based instruction, personal tutors, summer school classes, and sessions that combine academics with counseling at accessible locations in the community.

In addition to decisions on where and how to conduct classes, mandated functions need to be reassessed with awareness of the changed environment. For example, in 2007, the state of Louisiana suspended its Graduation Exit Exams because of the impact of Hurricane Katrina. In 1999, Advanced Placement exams were rescheduled for Columbine students. At Virginia Tech and other schools that have experienced a shooting in the springtime, graduation plans were adjusted, and students given options for completing the semester.

To provide home-based, online, or personalized instruction, it is likely that additional substitute instructors and staff will be needed. With an up-to-date list of retirees and other faculty who could step-in, a district or university can more efficiently address potential staffing needs. Crisis response agencies can help by identifying practitioners qualified to provide trauma counseling for children, adolescents, and adults; family-therapy counselors; counselors with expertise treating individuals with physical or cognitive impairment; and individuals to help with transportation and the many other needs that develop in the aftermath. Parents can help by building strong relationships with families of other students at the school and by involving their K-12 children in community activities—scouts, youth groups, service projects, intramural sports, and other pastimes—that provide opportunities to work in small groups and with other adults. This builds a broader "safety net" should disaster occur.

While it is likely that leaders will need to hire additional teaching, clerical, and support staff, it is almost certain that they will need to screen out volunteers who flock to the scene. After a disaster, many good-hearted but often ill-advised souls want to help and show up to lend a hand. One administrator shared with me that he was overwhelmed by the mass of people who came to his school: "So many wanted to be associated with the tragedy— some to help out, some purely for personal gain—that it was hard to filter out even those who were well-meaning. Everything takes so much energy, and energy is hard to come by when the world shatters right in front of you." He advocated that additional clerical support be hired simply to deal with offers of assistance. Recovery plans should also include strategies for productively utilizing volunteers who do have something to contribute. One program director was grateful that the administrative support his district authorized took over the routine running of his program, so he could focus on the added demands related to the tragedy.

Additional flexibility will be needed for a while, not just at the resumption of classes. For example, at Columbine the following year an increased number of school assemblies and motivational speakers who visited the school elicited different reactions. Some appreciated the chance for everyone to get together, feeling that it "knit us together and built a sense of community." Others avoided such events: "There were more assemblies than usual. I didn't go—I heard that speakers did give strategies that were helpful, but I knew that just being crowded into the auditorium did not appeal to me. I did not want to be trapped in there with so many people."

The existence of such differences highlights the need for alternatives that take into account the individual nature of the trauma response. For example, students who become anxious in crowds could be given options for alternative activities. If the school doesn't provide those options, students are likely to create their own. As one student admitted, "I hid out in a classroom with a teacher who hated those assemblies too!"

Feelings of anxiety in closed spaces are a natural reaction for students who have been trapped for hours in closets or classrooms. Individuals from schools that experienced a shooting indicated that they routinely look for exits whenever they walk into a room. Everyone needs to know what can be done to escape a situation. If escape is not an option, they need to know how to secure a door against intruders, how to call for help, and what will be done to assist them.

### EMPOWERMENT AND SAFETY

A disaster brings chaos and threat, and reclaiming a sense of personal safety is a major challenge for survivors. A traumatized teacher or faculty member may struggle with tasks that used to be easy. A fearful student may find it difficult to learn new material. However, a routine that is predictable, a safe environment that is supportive and flexible, and access to reliable and timely information can reduce anxiety and contribute to the recovery process.

While a return to class work is a positive step on the road to normal, achieving a sense of continuity requires that what transpires during the day "connects" to what was known and done before. For example, devoting excessive time to "healing" activities for days on end—to the exclusion of functions more commonly associated with school—emphasizes just how different things are. At some point, school needs to be school, even if expectations and academic assignments have changed. Keeping students in the equivalent of academic limbo impairs an authentic return to school functioning. Allowing students to resume projects they had started before the tragedy can help them reclaim a sense of purpose, and with the option to either resume an earlier project or begin a new one, students are empowered to choose for themselves. One student lamented that when she returned after a tragedy, she found that plans for publishing a student magazine had been "thrown out the window" and replaced by a "healing" project:

> I wanted to finish what we had started before—all of the writing that we had been working on. It was really important to us, but we were never allowed to finish it because we were sidetracked to a different project. That made me angry. It's not just starting new things that are intended to "help us heal" but having an opportunity to finish old things.

In an encounter with disaster, people are victims of circumstance. They did not choose their experience and often have little power to change the outcome. Victims are subjected to forces outside of their control, and as soon as possible after a traumatic event, survivors need to regain some measure of self-empowerment. At the same time, it is important for schools to balance expectations in a way that acknowledges the student's psychological distress while not enabling or condoning potentially destructive behavior. An orderly school environment contributes to stability, confidence, and safety, yet rules need to be applied with an awareness of trauma's impact. Children and teens may regress to behaviors that they have outgrown and seem less mature than they had been before the tragedy. Rules for school and campus behavior may need to be renegotiated, with students being given a voice in how to handle disruptions and the possible consequences. Students may need an option to request a "time out" when feeling overwhelmed. Students noted that heavy-handed interventions imposed in the name of safety after a shooting did little to reduce anxiety. In fact, many said they thought additional security measures would not have prevented the rampage and were just being done "for show."

Students aren't alone in a need to feel that their voices are heard. Teachers also expressed frustration when decisions were made at the district level without having an opportunity to provide their input. Faculty wanted a voice in making decisions about things that affected them and sometimes felt their professional identity and abilities were being disregarded. Being asked to help negotiate compromises, generate alternative ways to achieve school mandates, or brainstorm solutions to problems would have conveyed a much-needed sense of validation. The essential element is that teachers, students, and staff are respected and listened to. Their voices, their opinions, need to be solicited and heard.

## TALKING AND LISTENING

Genuine, clear communication is at the heart of effective educational practice. Following trauma, it has even greater significance, for the simple process of listening can be life-sustaining. The parent of a child killed in a school shooting gave some good advice for communicating with survivors in the aftermath: "Kids develop their own vocabulary to talk about it. They have to face it. Parents and teachers need to learn what *their* vocabulary is in order to be able to communicate with them."

A guidance counselor who contributed to this chapter said that what helped students and teachers most was for people to just listen to them. "If you encourage people to talk to you, it is important that you let them talk freely; just listen without interruption. Don't try to argue them out of their pain or anger, but let them know if their words, actions, or threats concern you." Parents find it hard to listen to their children that way: "The kids wanted to protect their parents because they knew how traumatized their parents were, so they hid things from them because they knew that their problems were hurting their parents. Parents really needed training to know what to do, not just lists of symptoms to look for."

Listening says that you care, that you can be trusted to receive troubling information without judging or attempting to *fix* the pain with conventional wisdom that no longer applies. Life has changed for all within the trauma membrane, and this makes platitudes seem especially naïve and blatantly false.

A student commented that caring people make the difference in every situation and that students and school personnel need a chance to get together with people who understand them: "I would make sure that there are many opportunities for students to get together with other students, teachers to get together with other teachers, support staff with support staff, and sometimes everyone together." Having a chance to talk with people who were in the same place you were at the time of the tragedy is especially beneficial, for it helps to sort out what happened and to "test out" memories with people who had shared that space and time.

Talking with friends, family, classmates, or colleagues can be reassuring, yet some faculty candidly shared that their spouse wasn't able to "be there" for them in the way they had always been in the past. The extreme nature of the traumatic experience puts stress on relationships. Family members also need support and opportunities to learn about normal trauma reactions and to get together with other families.

When a tragedy affects one family member, it affects the whole family. The impacts on the siblings, spouses, parents, and others are often overlooked, yet the anxiety caused by the experience is likely to echo for years. Several teachers and administrators reported that the stress of the experience contributed to their marriage ending in divorce. The wife of a teacher stated that she realized she had to accept that her husband was not the same person he had been prior to the attack. As she described it, she needed to grieve the loss of the person he had been in order to find the person he had become—and fall in love with him anew. Students reported deep and painful disruption in relationships with parents and siblings: "Everyone wanted to help me, but no one really understood what I was going through. My parents just infuriated me—we used to be really close, but now every time I see them I get angry." Knowing in advance what people are going to go through can help family members understand that everyone has been affected and that moving beyond trauma will take time.

A major concern following a traumatic school event is how to provide ongoing support for students who graduate, transfer, drop out, or move out of the area. Students and staff who soon leave the school or campus often lose the personal connection with those who shared their experience and the nurturing support from community members. They also may lose access to the mental health services that are offered to those who remain in the area. The growth of social media networks has facilitated the giving and receiving of support for those at a distance. Communities are no longer tightly bound by physical proximity. Technology resources can be used to provide social support through online access to friends and counselors, web-based groups (e.g., Facebook, LinkedIn), webinars, and other tools, thereby lessening some of the challenges experienced at a distance.

### DEALING WITH TRAUMA PROACTIVELY

Long *before* a disaster happens, schools can help students develop the conceptual understanding of traumatic experience, a vocabulary for discussing the occurrence, what can be done to minimize or escape danger, and strategies for rebuilding and recovery. Teachers, faculty, and counselors emphasized the need for grade-appropriate curricula to prepare *all* students for crisis and "show them how to survive if one should hit." They said they wanted a chance to learn about trauma and disaster readiness as well.

Introducing any new curriculum into a K-12 standards-driven academic environment will require either developing content standards for crisis or integrating learning activities within existing subject areas, potentially with a focus on geographic location and relative risk. A study of the science and frequency of natural disasters, for example, would be a reasonable expectation in locations likely to experience those events. Disaster preparedness, response, and problem solving in times of crisis are appropriate considerations. Helping students conceptualize critical events in history as *human* experience could enhance their awareness that disasters happen to real people; for example, the destruction of Pompeii can be taught as a human event, not just a geologic occurrence or an archeological oddity. Such broad concepts can be addressed in a literature curriculum through nonfiction and fact-based fiction that show human challenge and resilience in facing a natural disaster, violence, or warfare. Social science courses can explore the consequences of terrorism and diaspora for civilian populations. By providing opportunities to learn about crises, understand what is necessary for survival, and verbalize thoughts and fears, schools help prepare students to survive if a disaster should hit.

Similarly, curriculum in interpersonal relationships, communication, and problem solving can foster development of survival skills. A Columbine graduate, who is now a teacher, says he gives his students "life problems" to solve on a daily basis. He wants them to be prepared to make life-saving decisions and uses materials like *Kelso's Choice Conflict Management Program* (http://kelsoschoice.com) to help his elementary students learn to think critically and resolve disagreements, with or without the intervention of adults.

Those who personally experienced a catastrophic event, without exception, advocated adding trauma studies to the standard curriculum, emphasizing that schools and universities can "go a long way to help prepare individuals for any occurrence of trauma and loss in their life." Instead of encouraging complacency and denial by neglecting the subject, educators can prepare themselves and their students for life-challenging occurrences. Understanding common reactions to trauma, including the neurological, physical, psychological, and socio-emotional effects can help them in a personal encounter with disaster and prepare them to interact with those who have survived traumatic loss.

As a student described it, "Loss of innocence, loss of expectations, loss of happiness—it feels like you've been thrown into a dream and nothing is real anymore." In the aftermath, counselors and families attempt to help survivors deal with the fatigue of sorrow, fear, and depression, yet they are dealing with situations for which few have prepared. It is not widely known, for example, that trauma is likely to manifest not only as empathy-arousing emotions of sorrow and grief but also as vicious displays of anger and rage, which are unpleasant to countenance and often generate anger in response. As one student so poignantly described it,

> I was really angry, not just at [the killers] but at the media and basically every-body. Everyone said it's okay to be sad. It's okay to have bad days or to have memory problems. It's okay to grieve—but nobody ever told us that it is okay to be angry, and anger was a big part of it. I was so angry that I was pushing my parents away. I was just fighting everything in my life. If I had been able to deal with anger in a better way, things might have been easier. No one said that it is okay and no one gave me any help with dealing with the anger. It was as if they could deal with sorrow but anger is such an ugly emotion that nobody knew what to do except say things like, *calm down*, and *don't be so angry*. That didn't help at all.

In recovering from a trauma, people need to understand and come to terms with their relationship to the event. A parent reminded me of the "unique wound and fear among young people who have been traumatized in these situations—the wound of stigma. Young people are particularly vulnerable to it. 'Now I am different. People will see me that way and I will never fit in again.'" She went on to say that educators are presented with a "wonderful/terrible opportunity to show kids that they are special and they can do special things because of this experience." Educators can help in this process by creating new traditions, finding ways to tap the experience, and giving students opportunities to express themselves and create a new and powerful insight from their experience. In this way, students can learn that because of their experience they can be *more* than they were instead of less.

Some who have experienced trauma openly speak of their fight with depression and PTSD, revealing early thoughts of suicide and substance abuse. However, many choose not to talk, not wanting to forever be associated with a tragedy or fearing the stigma of mental illness. A victim's advocate shared with me that some parents were so concerned about the stigma of mental illness that they wouldn't take their children for therapy. "In other families, where there had been some dysfunction to begin with, parents refused to take their kids to counseling for fear of what secrets they might reveal. It makes for a very difficult situation."

An appropriate curriculum along with a public awareness initiative related to disaster readiness and traumatic response can help reduce risk, facilitate recovery, and begin to minimize the stigma of mental health issues. An educational program focused on how to deal with extraordinary events would be most beneficial *before* a trauma.

## HEALING ACROSS THE CONTENT AREAS

When a school or university has been exposed to a life-threatening circumstance, developing and scheduling instruction to support learning is difficult. After a disaster, subsequent disruptions often create a shifting landscape. Classroom assignments must be adjusted to meet the altered learning needs and abilities of the students, and accommodate the changed environment around them, and may take on a therapeutic quality.

A variety of options for assignments offers a respite from highly energized and stressful engagement. Redesigning lesson plans to include more small group work, breaking down large units of study into smaller components, and providing students with more opportunities for self-expression through a creative medium of their choice—journaling, storytelling, art, dance, drama, photography, creative writing—are simple approaches that can facilitate learning while students are trying to process their feelings about what happened.

For journaling or text-based expressions, students whose native language isn't English should be encouraged to process their feelings in their home language. Culture, religion, social norms, ethnicity, and family traditions influence a person's understanding of and response to tragedy as does gender and developmental level. People knowledgeable about the cultural and social heritage of the students can help teachers anticipate potential reactions. Special learning needs must still be met, and it is important to remember that linear-sequential thinkers might not respond well to "artsy" or creative units that seem to be going nowhere.

Monette Park, a social worker at Columbine, observed that after the shootings many teachers would have preferred to return to their standard teaching methods yet modified their instruction to meet student needs. She noted that people tend to return to the familiar after a situation that is stressful, confusing, or overwhelming.

> After the shootings, all of us were experiencing the shock of discovery that our lives were changed forever and the sadness that things would never again be the way they were before. Accepting this dramatic change of identity—as individuals, as Columbine, as parents, as a community—was very difficult. As a result, some of the teachers wanted to return to the known ways of doing things. This was hard for some of the kids.

However, as Park points out, "It needs to be recognized that the school should not be an adolescent treatment program with an educational component. It needed to be a setting of education with a mental health support." Students were "normed" by being students—by learning and activity and academic rigor, to the extent possible. She observed that the most powerful growth toward wellness came from teachers doing the best they could to make the kindest possible environment for students. The only normal became the process of "doing one's job" and doing it as well as possible:

> The sensitivity that staff initially displayed toward student needs was still happening, a willingness to cut some slack if a student started to struggle. It became almost a core belief by most staff that any kid would "snap back" as soon as they were able. Then they'd start to achieve again at a reasonable level. Students and staff did not want that one experience to become the sum total of who they were and where they fit in the world.

To help ease a return, many teachers spent the first days back in the comfort of review. It helped students to get back to what they knew. In the following year, many teachers accommodated their students' difficulty with memorization by allowing them to use a "cheat" card—an index card filled with information such as chemical formulas, mathematic equations, historic dates and places, and other facts they would need for tests.

Curricular support for Columbine came from a variety of sources. Because of a long-standing relationship with the Colorado Artists-in-Residence Program, the school received a grant from the National Endowment for the Arts that funded artists to visit the school and help incorporate arts and creativity in diverse subject areas. As art teacher Barbara Hirokawa described it,

> In the first days back, teachers were at a loss for how and what to teach so artists came to the classes that requested them and worked with their students. A sculptor came to the chemistry class and had students make whimsical sculptures of molecular structures that were a play on words. The physics classes made tiles. The special education teachers did a variety of projects. Students and teachers talked, played, wrote, acted out in theater, danced, pounded on drums, created sculptures, and tentatively began letting some of the emotions out.

Working with the Artists-in-Residence, teachers introduced creative activities into different content areas, including English, chemistry, social studies, physical education, physics, and, of course, the arts. As Hirokawa explained, "Creative expression allows people to communicate things that cannot be put into words, giving voice to express the depths of grief and horror and a means to speak the unspeakable, and by doing that to begin to heal."

Instrumental music and choral teachers helped students find solace in musical expression. As soon as possible, students retrieved their instruments and sheet music and were practicing for orchestral and choral concerts for the school and community. Self expression, in whatever form, offered an outlet.

Literature opened an opportunity for students to talk about the pain of surviving in the midst of tragedy. An insightful English teacher assigned her classes to read the novel *Ordinary People* and facilitated a dialogue on survivor's guilt and suicide. Another equally mindful teacher replaced *Hamlet*, Shakespeare's tragedy of betrayal and suicide, with one of the Bard's lighthearted comedies. Both were creative and caring teachers who chose different approaches, but each was purposeful. Each weighed the impact and the importance of classroom activities in terms of achieving curriculum requirements and meeting the changed needs of students.

Students appreciated the variety of activities the teachers provided. Some told of benefitting from a teacher's introducing them to the basics of yoga breathing to help them center and calm themselves to relieve stress and anxiety attacks. One student found particular solace through a photography class, which, she said, gave her the power to express her grief without words. Another said that a philosophy class gave her the opportunity to talk and deal with feelings about death in an enlightening way. One took a self-defense class, which helped her develop a sense of well-being, knowing that, if confronted again, she was better prepared to respond:

> Taking the defense class gave me a way to deal with the anger and helplessness I felt and to reclaim a sense of power. For people who have been a victim of crime, whether in a school shooting, or on a city street, or in their home, a self-defense class gives the skills to fight off an attack and that will actually help in processing the trauma they experienced in the first place.

### HEALING BY HELPING

Gerda Weidman Klein and her husband Kurt first visited Columbine in January 2000 to offer her support for the grieving community. As a young teen, Gerda lost her family to the Holocaust. When she stood before the auditorium filled with students, teachers, and families, she began her talk by simply saying, "I understand." With those two words, she connected with everyone in the room, for she had survived a trauma that far exceeded their own, and yet here she was, offering a message of hope, an example of survival, and a model of using one's own survival to help others.

Gerda and Kurt returned to the school several times over the years, continuing to meet with the students and faculty, helping to raise funds for the memorial, and becoming cherished members of the school community, as evidenced by the fact that every Columbine student, educator, and parent interviewed for this book cited Gerda as a positive force in the aftermath. Taking her lessons to heart, students begin a service organization and founded "The Heart of Columbine" as a way of reaching out to help others. They agreed that each year, the organization would choose a different effort to support. The first year, they committed to help fight hunger, an issue that was particularly important to Gerda and Kurt.

Much as the I Love U Guys Foundation benefits school and community in Bailey and elsewhere, engagement in service projects offered Columbine students a significant opportunity to achieve a sense of purpose outside of self, build a connection to the greater community, gain an appreciation for their own capacity to assist others, and learn about larger issues beyond their own experience of tragedy. Many survivors have gone on to careers in the helping professions—teaching, counseling, medicine, disaster prevention, and search and rescue, to name a few.

## ADVICE FROM PARENTS

My interest in finding ways to help students and educators began with my own experience in the aftermath. As I researched the effects of the shooting, I interviewed other parents and asked their recommendations for what could be helpful in similar situations elsewhere. The Columbine shootings, of course, pale in comparison to many other tragedies of natural or man-made origin. The disruption that follows a natural disaster or terrorist attack creates challenges at a different level of magnitude. However, pain is pain, and it is always possible to learn from another's experience.

Parents in the original research and in subsequent interviews described their struggle with the ongoing nature of the trauma. They described their appreciation for the contributions of teachers who returned the following year, the importance of family, and the need for proactive educational planning and mental health services throughout the educational system. Their stories reveal common impacts of primary as well as secondary exposure to lethal violence.

Parents struggled to reconcile to the fact that their community was no longer just an average, anonymous, American suburb. It had joined the list of locations made famous for being the site of a rampage school shooting. The disbelief reveals the common assumption that "these things" happen elsewhere, but "certainly not here." Yet, Columbine was no more exempt from violence than were other unsuspecting communities. Undeniably, violence happened here, prompting one parent to observe, "We are all vulnerable, just hanging by a thread."

Parents discussed one particularly painful point that can easily be avoided in other settings. Since many parents saw their children struggle with trauma response and PTSD, the language that was being used to describe the event caused dismay. Early on in the tragedy, the term *victim* was used to denote those who died in the attack, and the term *survivor* was used to signify those who had been shot but recovered from their wounds. This left no term to describe those who had not been physically injured but who were psychological victims of this crime. Over the years, the issue has resolved itself, with reference now being made to anyone who experienced the tragedy as a *survivor*. However, this situation demonstrates the importance of care in choosing language to describe an event.

Some parents whose children were killed that day still express outrage at law enforcement, the school, and parents of the shooters. Others who lost their children have become spokespersons for service projects, including youth ministry, school safety, and character development. Some have turned to political activism, such as lobbying to prohibit sale of guns to minors. Most, however, have resumed a quiet life, content to be out of the public eye.

Many parents whose children survived the assault said that people sometimes still treat them as if they have been permanently damaged by their association with the tragedy, as if they were "psychological basket cases." One parent said that whenever she meets someone for the first time she doesn't tell them she's from Columbine, "because people think you're broken or that there was something wrong with you or your community for this to happen to you."

Several parents reminded me that in many cases, measures to help make students feel safe in returning to the building actually increased their anxiety and distress. For example, requiring students to wear ID tags when they had never worn them before, actually added to the anxiety as students considered how unlikely the new policy would be in deterring future violence.

A police officer who was on the scene at Columbine observed, "All the security in the world would never have stopped those two." He felt that settling on bullying as the cause of a massacre painted the gunmen as the aggrieved rather than as the aggressors. He also advocated that mental health be attended to throughout a student's school years as a way of violence prevention, not just to promote healing after a tragedy strikes.

In the aftermath, parents were uncertain about how to parent their traumatized teens. This unfamiliar territory had major implications for their child's future, and they needed solid information and guidance delivered in a way that they could understand and process, not just the sort of lists that were freely distributed. They were dealing with their own traumas and struggling to understand what the lists meant, what they needed to do, or how they could help.

In addition to informal social support and professional services provided through mental health agencies, school district, and private organizations, some parents appreciated support provided by area churches. From the very first evening, the churches in the Columbine area became places for reconnecting. Local pastors had agreed that the tragedy was not a time to recruit converts, and it was a time to offer comfort and gentle support to all who came through the door. Even parents who had no affiliation with the churches shared stories of going to services and finding encouragement.

In the aftermath of traumatic events, people try to find meaning, and in the process of reconciling to their experience, they look for positive outcomes. Many felt that dealing with their experience moved them to a higher level of consciousness, a sense of expanded awareness and a greater capacity for empathy. Some voiced a feeling of increased connection: "I never felt so connected in my whole life. In the loss, so much was given." As one parent described it,

> You had to move outside of yourself, outside of your own pain and deal with a whole community of pain on so many different levels. It has made you a more complete person. But they think you're damaged goods because they think you will never return to the place you used to be. But that's not the point. The point is to move to another level of experience, where you process all that you used to be and all that you are now into who you want to be. And that's the point with anybody who goes through a loss in their life.

### REFLECTIONS

The insights shared in this chapter suggest several common themes—no one expects to experience a disaster; people respond to traumatic events differently; needs and abilities are likely to change; adaptations are required; and the aftermath lasts for a long time. Compounding the grief that is felt for those killed and injured, after a disaster many are burdened with an overriding sense of loss—loss of a sense of safety, loss of a sense of place, loss of connection to others, and loss of confidence in the future. Previously held beliefs and assumptions about the world have been shattered, and it becomes necessary to construct a new worldview that acknowledges the experience. One of the Columbine parents shared with me her perspective on life after the shootings. Trauma changes things, but life endures.

> If someone asked me,
> When we will get over this, I'd say,
> I've heard this description:
> We all have a house that we live in,
> In our house we have different rooms.

Columbine is one of the rooms in our house.
And it's always going to be in our house and
It's not going to go away.

It doesn't mean that we have to go in that room every day,
But it's part of who we are now.
It's always there and how often you choose to go in—
How often you need to go in—
How often you do go in—
So be it.

Someone who doesn't have that house in their life,
Someone who doesn't have that room in their house,
Probably can't relate to it.
They just don't know.

(Mears, 2005, p. 86)

## REFERENCES

Janoff-Bulman, R. (1999). Rebuilding shattered assumptions after traumatic life events: Coping processes and outcomes. In C. R. Snyder (Ed.), *Coping: The psychology of what works* (pp. 305–323). New York: Oxford University Press.

Mears, C. L. (2005). Experiences of Columbine parents: Finding a way to tomorrow. Doctoral dissertation, University of Denver. ProQuest UMI: AAT 3161558.

Mears, C. L. (2009). *Interviewing for education and social science research: The gateway approach.* New York: Palgrave Macmillan.

Zoba, W. M. (April 2, 2001). Columbine. *Christianity Today*, 55–59.

**WHAT'S NEXT?**

This chapter concludes the individual accounts of resuming school after a crisis and the advice for what helps and what hurts recovery in the aftermath. Section Three provides a synthesis of common themes and insights taken from the experience of school-related trauma and an annotated list of materials and resources that can help with crisis prevention, response, and recovery.

# SECTION THREE

# PUTTING PAIN TO WORK

CHAPTER 13

# SO NOW WHAT?

THROUGH THE STORIES YOU HAVE JUST READ, YOU NOW know more about the world of trauma. It can no longer be a remote, academic awareness, for you have met, by virtue of their words, real people who were thrown into a world of chaos and disruption. The authors are people just like you. They come from rural communities, metropolitan centers, suburban schools, prestigious universities—a growing list of locations in which lives have been taken and futures changed. It may be hard to imagine any of these events occurring in your school or community. Remember, the authors felt the same—*before*.

This text is my effort to demystify the experience of trauma, to give you a view from within the membrane that separates those who have lived a life-changing circumstance from those who have not. I hope that you will recognize this as a call to action, because with awareness comes responsibility. The experiences that authors have shared, the problems they faced, and the solutions they found are lessons to learn and put to use. Knowing what has helped others should be instructive. That is the *so-now-what* of this book.

Law enforcement, government agencies, first responders, and counselors prepare for crisis, hoping it never happens but knowing that at some point their preparation will be put to the test. The role of educators is to facilitate learning for every student, and that requires preparing for the possibility of crisis too. If a tragedy strikes a community—or a single student—teachers have the same school responsibilities that they had before. But they and their students now have new and very critical needs that must be addressed.

No universal prescription can be written for reclaiming a school after trauma. The "appropriate" response depends on the event, the individuals involved, and the community itself. The type and scale of the crisis, along with the availability of resources and support, will combine with cultural, religious, and geographical differences to make each case unique. No matter how comprehensive a planning effort, it is simply not possible to predict and plan for every contingency. However, as the authors have shown, some aftereffects can be anticipated, as can the need for certain types of services. Readiness to respond can improve the outcome. Educators can help students and families learn about disasters as a part of human experience, understand the effects of exposure to trauma, and build an appreciation for individual resilience and community strengths. By helping students develop critical thinking and problem-solving skills, educators can prepare them for the challenges that life presents. Students can benefit from explicit instruction about what to do in an emergency, how to survive, how to reconnect with family, how to access services, and how to assist others.

The disruption that follows when a school has been terrorized by violence, accident, or natural disaster makes the world seem threatening and strange. The most immediate need after any traumatic circumstance is the need to feel safe and to reconnect with something resembling *normal.* People need to know what is being done to protect them from further violation. Effective communication that provides accurate and timely information is essential. The media can be turned into an ally in achieving this goal.

Returning to a reasonable schedule, to the extent possible, can be comforting. Whether it is the same school schedule as before or something quite different, the routine of classes can introduce a much-appreciated sense of order and predictability. Welcoming students, staff, and families into the planning and decision-making process conveys a message of respect and says that their opinions matter. Acting in the best interest of *all* who have been affected can avoid fueling resentments and controversy.

A community-wide event introduces the stark reality that tragedy has happened "here." Individuals and the community as a whole may struggle to build a new sense of identity to incorporate this experience. Blending what was known and valued before with the transformative effects of the tragedy can lead to a positive, new identity. By honoring the losses, celebrating the strengths, and respecting the many ways that individuals process grief and pain, educators can help their school community come together to support each other as they move toward this essential aspect of recovery. Providing a variety of opportunities for people to get together with friends, family, and colleagues to reflect and share experiences can help bridge the isolation and disorientation that often accompany trauma and can foster a sense of shared purpose and understanding. Supporting service projects for survivors to be involved in assisting others can help them regain a feeling of empowerment. And, especially important, tending to one's own recovery can increase the potential for being able to help others.

Stories of traumatic experience are unsettling, yet they are essential. Knowing what to expect will not make the challenge go away. However, learning from past events can help prepare you for circumstances you may someday encounter. Each of the authors has shared a unique experience and perspective to provide you with a variety of insights, reflections, and approaches that can advance the successful return to school and attend to the needs of students and school personnel. While schools are places of teaching and learning, they can also be places of support and healing.

If your school hasn't considered the possibility of disaster, or if insufficient attention is given to the needs of students exhibiting traumatic stress, it is time to begin the conversation. Work within your school community, perhaps using this book as a starting point for discussion. Develop your own capacity for meeting the special needs of students after trauma, and share what you learn with others.

The well-being of students, faculty, staff, and all within a community depends on those who are willing to prepare for the unthinkable. I encourage you to take on this significant challenge so the lessons learned by others will not be wasted.

# CHAPTER 14

# RESOURCES

THE FOLLOWING LIST REPRESENTS ONLY A SMALL SAMPLE OF the many helpful organizations and resources, research, guidebooks, curricula, handouts, and other materials that can assist educators and institutions in preparing for disaster and in facilitating recovery in the aftermath. This list is not meant to be all-inclusive, nor is it intended to endorse any program or practice for your school or circumstance.

Web addresses for resources are subject to change, but names of agencies and organizations generally do not. If you cannot readily find an item listed below, go to the home page for the publishing organization and use their search engine to locate the material.

## DISASTER READINESS, RESPONSE, AND RECOVERY

*Child Trauma Academy* offers a variety of articles, research, materials, and courses, including *Surviving Childhood: An Introduction to the Impact of Trauma.* Free download, www.childtraumaacademy.com and www.childtraumaacademy.org

*Campus Attacks: Targeted Violence Affecting Institutions of Higher Education* addresses prevalence, prevention, and response to targeted attacks on college/university campuses, compiles findings of a collaborative effort by U.S. Secret Service, Department of Homeland Security, Office of Safe and Drug-Free Schools, Department of Education, Department of Justice, and Federal Bureau of Investigation. Free download, www.fbi.gov/stats-services/publications/campus-attacks/campus-attacks-pdf/

*Helping Children and Adolescents Cope with Violence and Disasters,* a guide for parents of children exposed to traumatic events, is available through the National Institute of Mental Health. Free download, www.aap.org/disasters/pdf/helping-children-and-adolescents.pdf

*How Schools Can Help Students Recover from Traumatic Experiences: A Tool Kit for Supporting Long-Term Recovery* reviews programs and resources for helping students recover from disaster. Free download, the RAND Corporation, www.rand.org/pubs/technical_reports/TR413/

*National Association of School Psychologists* provides resources, fact sheets, and information for educators, parents, and community members on such topics as resilience, psychological assessments, crisis, bereavement, depression, suicide prevention, memorializing loss after a traumatic event, culturally responsive practice, military deployment, economic hardship, and related topics, including *Positive Behavior Support* (PBS), referred to in Chapter 5. Free download, www.nasponline.org/

*National Center for School Crisis and Bereavement* offers materials and support for educators, parents, caregivers, and disaster relief personnel assisting children and youth touched by crisis, including guidelines for helping students deal with instances of suicide and death in the school community. Free download, www.cincinnatichildrens.org/svc/alpha/s/school-crisis/

*National Child Traumatic Stress Network* provides an extensive array of information on multiple types of trauma and its effects and offers helpful materials, including *The Child Trauma Toolkit for Educators*. Free download, www.nctsnet.org/

*National Institute of Mental Health: Coping with Traumatic Events* offers a wide variety of research-based information and evidence-based practice, designed to help children, adolescents, and adults cope with traumatic events. Free download, www.nimh.nih.gov/health/topics/coping-with-traumatic-events/index.shtml

*Psychological Trauma of Crime Victimization* provides information on physical and psychological reactions associated with the trauma of crime victimization, with relevance to other traumatic circumstances as well, including cyberbullying and disaster. Free download, National Organization for Victim Assistance, www.trynova.org

*School Crisis Guide: Help and Healing in a Time of Crisis*, provided by the National Education Association Health Information Network, offers health and safety information, tip sheets, tools, and templates to assist school leaders before, during, and after a crisis. Free download, www.neahin.org/educator-resources/school-crisis-guide.html

*Tips for Helping Students Recovering from Traumatic Events*, published by the U.S. Department of Education, includes practical information for educators and parents helping those exposed to a traumatic event. Free download, www.ed.gov/parents/academic/help/recovering/part.html

*American Red Cross* has created a variety of materials and resources on disaster preparedness and recovery, including lesson plans designed to assist educators in helping students cope with disaster. Free download, www.redcross.org/pubs/

*Readiness and Emergency Management for Schools*, established by the U.S. Department of Education, Office of Safe and Drug-Free Schools, provides a wealth of guidance, interactive training materials, and other resources to support the implementation of National Incident Management System (NIMS) on school and college/university campuses. Free download, http://rems.ed.gov/

*School-Based Mental Health: An Empirical Guide for Decision-Makers* provides practical information and advice for those engaged in developing and implementing effective mental health services in the school setting. Free download, University of South Florida, Research and Training Center for Children's Health, http://rtckids.fmhi.usf.edu

*I Love U Guys Foundation,* created by John-Michael Keyes and Ellen Stoddard-Keyes after the death of their daughter in a school shooting (Chapter 8), provides training guides for students and staff on response to crisis in a school, including the evidence-based *Standard Response Protocol (SRP)*. Free download, http://iloveuguys.org/

*Resilience in Children and Teens: A Guide for Parents and Teachers*, a publication of the American Psychological Association, includes valuable suggestions for assisting children in building resilience. Free download, www.apahelpcenter.org/dl/resilience_guide-for_parents_and_teachers.pdf

*Teen Action Toolkit: Building a Youth-led Response to Teen Victimization* is a hands-on guide to help build youth resilience and strengths. Free download, National Center for Victims of Crime, www.ncvc.org/

*Tips for Talking about Disasters,* from U.S. Substance Abuse and Mental Health Services Administration (SAMHSA), provides information and classroom activities for educators on marking anniversaries of disaster, questions to encourage children to talk about their experiences and responses after a disaster, ideas for helping children express themselves through art and other means, guidance on the role of culture and ethnic background, measures for reducing anxiety and stress, and self-care. Free download, http://mentalhealth.samhsa.gov/

*U.S. Department of Education Office of Safe and Drug Free Schools* offers an array of high quality resources on school-based trauma intervention, violence prevention, mental health, and suicide prevention. Free download, www2.ed.gov/about/offices/list/osdfs/index.html

*U.S. Department of Education* offers a variety of resources for educators on disaster preparedness and recovery, including *Action Guide for Emergency Management at Institutions of Higher Education* and *Tips for Helping Students Recovering from Traumatic Events*. Free download, http://edpubs.ed.gov/

*U.S. Federal Emergency Management Agency* (FEMA) is a source for many useful tools and services to assist in preparation for and recovery from crisis, including *Recovering from Disaster*. Free download, www.fema.gov/rebuild/recover/after.shtm

*Impact of Mass Shootings on Survivors, Families, and Communities* in the *PTSD Research Quarterly, 18*(3), reviews research into the aftermath of shootings. Free download, National Center for PTSD, www.ptsd.va.gov/professional/newsletters/research-quarterly/V18N3.pdf

*Lessons Learned from School Crises and Emergencies*, a series, reviews actual school emergencies and crises (natural and man-made), critical actions, decisions, and events before, during, and after real incidents, and considers what worked and what did not. Free download, http://rems.ed.gov/index.php?page=publications_Lessons_Learned

*Lessons Learned: A Victim Assistance Perspective, 2006 Tragedy at Platte Canyon High School, Bailey, Colorado* provides an in-depth look at the victim assistance response to the Platte Canyon High School shooting (Chapter 8). Free download, http://dcj.state.co.us/ovp/Documents/OVP%20General/Bailey_Project_Lessons_Learned.pdf

*Mass Shootings at Virginia Tech: Report of the Review Panel* (Chapter 10) is available for free download, www.governor.virginia.gov/TempContent/techPanelReport.cfm). The *2009 Addendum to the report of the Review Panel* is accessible at http://scholar.lib.vt.edu/prevail/docs/April16ReportRev20100106.pdf

*The Report of the Columbine Review Commission* provides information on the attack on Columbine High School and subsequent response (Chapter 7). Free download, www.state.co.us/columbine/

*American Foundation for Suicide Prevention* provides materials to help educators understand and recognize suicidal behavior and to meet the needs of students struggling to cope with life events. In addition to resources on prevention, *After a Suicide: A Toolkit for Schools* offers guidance for schools in the aftermath of a suicide or other death among the school community. Free download, www.sprc.org/library/AfteraSuicideToolkitforSchools.pdf

*Promoting Mental Health and Preventing Suicide in College and University Settings* is available through the Suicide Prevention Resource Center's searchable database of resources on suicide, suicide prevention, and mental health materials. Free download, http://library.sprc.org/

*National Suicide Prevention Lifeline* offers wallet cards, guides, and other materials, many free for download, www.suicidepreventionlifeline.org/

*Center for the Study and Prevention of Violence* at the University of Colorado-Boulder provides a searchable database and clearinghouse of resources, research, and materials, including links to websites for state and national organizations, resources, and libraries. Free downloads, www.colorado.edu/cspv/index.html

*The Role of Mental Health Services in Promoting Safe and Secure Schools* explores the role of mental health services in developing and maintaining safe schools. Free download, Hamilton Fish Institute at George Washington University, www.hamfish.org/

*Striving to Reduce Youth Violence Everywhere (STRYVE)*, a national initiative led by the Centers for Disease Control and Prevention, takes a public health approach to preventing youth violence and provides evidence-based strategies, training materials, technical assistance, community workspaces, information, and tools. Free download, www.safeyouth.gov/Pages/Home.aspx

*The Stop Bullying Now! Campaign* provides a wealth of information for students, parents, educators, and others who want to understand bullying. Free download, www.stopbullying.gov/

*Safe Havens International Inc.,* a nonprofit school safety center, offers free resources to help schools create a safer and more effective learning environment, at www.safehavensinternational.org

*School Safety Partners* facilitates the building of partnerships for research and implementation on issues of school safety and incident response and recovery, at http://schoolsafetypartners.org/

# APPENDIX: THE GATEWAY APPROACH

For my research into the aftermath of the Columbine tragedy, I wanted to collect and share stories from within the experience so others could know the effects of a school shooting and be better prepared should such a tragedy strike their community. I conducted a series of three extended, in-depth interviews with each of six parents whose children were exposed to the tragedy, producing approximately six hundred pages of transcripts, rich with significance and meaning. As I searched for a way to preserve the parents' experiences in print, I came across a method for writing poetry from interview data (Richardson, 1992).

While my intent was not to create poetry, I needed to reduce the sheer magnitude of transcripts into narratives that could be shared. As an *insider,* I wanted a way to confirm that the parents' meaning was correctly represented, to communicate the intensity of *their* experience, and to give a context for their observations. To do this, I identified passages in the transcripts that were relevant to my research questions and removed transition or filler words, leaving only essential words and phrases. Then I organized these excerpts chronologically and thematically into a narrative that provided a gateway into the lives of the individuals and revealed a world of significance that would have been lost had I just summarized or paraphrased the stories.

By distilling the transcripts to their essence, the presentation became more real and powerful for readers, providing them with a deeper understanding of the complexities within the intense situation. Sharing stories through this type of evocative display strongly connected readers with the narrators. Since the data were presented with the original expressions intact, the findings emerged from the narrative, *telling* the story— instead of telling *about* the story.

I use the term *gateway* to describe the approach, since it opens a path to deeper connection and understanding, fostering a metaphorical "community" of experience. By providing readers direct access to the significance narrators attribute to events, those *outside* of the experience are better able to understand what it feels like to be *inside* the event. Similarly, the gateway approach provides a means for those *inside* to cross the boundaries and communicate with those *outside* who want to learn from the situation. In addition, narrators report that the interview and distillation process offers them an opportunity for deeper personal reflection. In that regard, the approach can be a pathway to increased self-understanding and empowerment. As one participant in the Columbine study noted:

> You should learn from [an experience]
> I've shared some things with you
> Whenever you share,
> You have a better understanding of things.

Complete Columbine narratives are included in my dissertation, *Experiences of Columbine Parents: Finding a Way to Tomorrow* (available through ProQuest UMI AAT 3161558). Detailed explanation of the gateway approach to research may be found in my text *Interviewing for Education and Social Science Research: The Gateway Approach* published by Palgrave Macmillan (2009).

### REFERENCES

Richardson, L. (1992). The consequences of poetic representation: Writing the other, rewriting the self. In C. Ellis, & M. G. Flaherty (Eds.), *Investigating subjectivity: Research on lived experience* (pp. 125–140). Thousand Oaks, CA: Sage.

# About the Editor

**Carolyn Lunsford Mears, Ph.D.,** received her doctorate in Educational Leadership and Policy Studies from the University of Denver. As a parent of a Columbine High School student exposed to the shootings, she conducted dissertation research into the impact of the tragedy on schools and families. Her research, *Experiences of Columbine Parents: Finding a Way to Tomorrow,* was recognized as the Outstanding Qualitative Dissertation of the Year 2005 by the American Educational Research Association (AERA). In response to requests for information about the distinctive research approach she developed for her dissertation, she authored *Interviewing for Education and Social Science Research: The Gateway Approach* (Palgrave Macmillan, 2009), which was selected as a finalist as AERA Book of the Year 2010.

Dr. Mears holds a research appointment and is adjunct faculty at the University of Denver's Morgridge College of Education, and is a member of the Trauma Certification Board of the Graduate School of Social Work. She has published numerous articles and presented to audiences in the United States, Europe, and Australia on such topics as trauma response and recovery after school shootings, safe school environments, leadership in times of crisis, and qualitative research.

# INDEX